ISBN: 979-8-9915202-2-5

Published By: Real2Reel Cineservices LLC

Daniel Rowland
7205 NE 275th St.
Battle Ground, WA 98604

Designed by: Levi Ness
Edited By: Ryan Fleming
Contributions by: Spencer Pierson

HOW TO RAISE
1,000
CHICKENS
FOR YOUR LOCAL
FARMERS MARKET

Learn to raise, slaughter, butcher, and sell chickens
from a 5-acre homestead, with just 2 folks.

WRITTEN BY
DANNY ROWLAND

CONTENTS

There are so many people that I owe acknowledgement to in the creation of this book. First off my family, both blood related and chosen. You have always been with me.

Mom, you always told me "You can do anything you want, as long as you put your mind to it." Thank you so much for those words and wisdom. It's not always an easy path to believe in oneself but so much easier when you have a cheerleader!

To Sharon for building up your "Oak Tree", nourishing my roots, trimming the branches that needed trimming…and planting a forest to grow tall in, you are the best and most gracious partner. I love you.

To my children for being my motivation in all things, I want you to have the best world and I'm going to do the best I know how, to make it happen for you. With all my heart and soul I will be your strength if you need it, branches to climb up to build your dreams, some shade and protection from the universal winds and rains that try to drench your spirit, and be the proof that you can do anything you want in this life and the next.

To Kjartan for being happy to see me, and Ulfredsheim for showing up and filling our wood shed after my back injury, we would have frozen that winter without you all.

To all of you who have bought our farm products and supported us from the get go, contributed to our fundraising, supported the EV Tractor conversion, supported our kids in their 4H endeavors, and all of our farm projects from chickens, to sheep. Thank you for believing in, and working with us, toward a better future for the world!

And last but not least, my awesome team and good friends Spencer, Ryan and Levi. Thanks so much for your professionalism, knowledge, and skills making this, without you this life adventure would never have made it to the page.

In awe of you all:

Danny

INTRODUCTION

When my wife Sharon and I first started planning out this book, there was one thought that kept going through my head: When did we finally make it as farmers? It turns out that the answer is a little more complicated than I would initially have thought.

At first glance, there are plenty of options to choose from. I could point to the moment we left the city behind and moved to my family's farm full time, or the moment when we officially took possession of our own land and opened Misty Frog Acres. Or if I wanted to be pragmatic about it, the moment we sold our first eggs at a local market was a big moment, and getting accepted to our first farmers market was huge for us.

But on the other hand, maybe I shouldn't focus on the successes. Part of being a farmer is overcoming your failures. And there have been plenty, some of our own making, some that were out of our control. There were diseases caused by animals cross-contaminating fields, and viruses that came across the ocean to wipe out millions of birds nationwide. Or maybe it was when I was laying on the ground with a life-changing back injury, as one of my dumb llamas smugly stood over me while chewing without a care in the world. Overcoming those events was a major part of our journey.

Then there are also the moments where you have to go off script and solve problems that are unique to farming, like having to fix a tractor or build your own coop. Farming can be tough enough at the best of times, but when something goes wrong and you have to improvise, that can make or break you. So, was that when we felt like we made it?

The truth is, it's probably much simpler than that. The moment we made it as farmers was the moment we decided to do it. All it really takes is commitment. And that's part of why we decided to write this book.

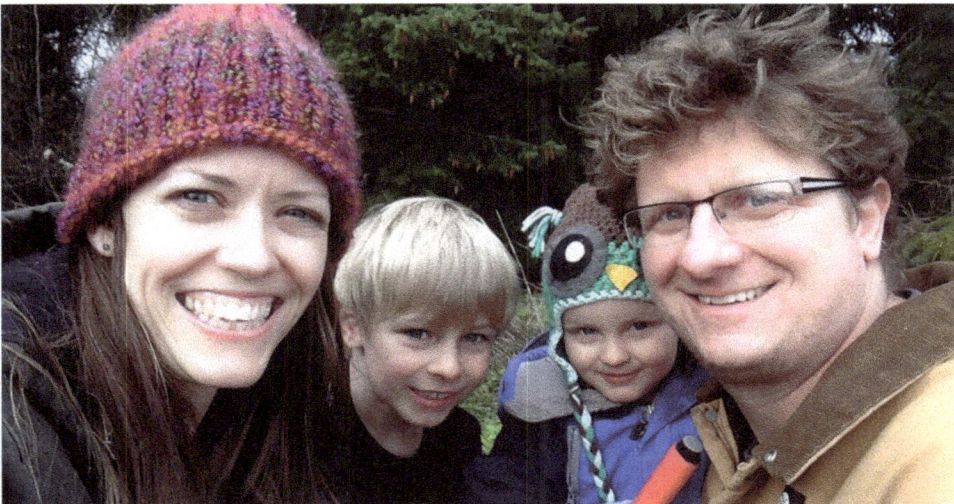

After more than a decade of living the farming life, with plenty of highs and lots of stumbles, we reached a point where we were finally comfortable and confident in our routine. We developed a process that works for us, and we think can work for others. It wasn't easy, and it wasn't always clear on the best way to proceed, but through trial and error we came up with a way for two people to raise 1,000 chickens for your local farmers market. Some of this will work for you, some of it won't, and that's fine. Most of what I did was just trying a bunch of stuff, then inventing new methods that sort of worked, followed by years of tweaking. That's part of farming.

But a lot of our knowledge comes from talking and reading the experiences of many more. So for us, writing this book is about adding to that communal database that anyone can use. At its heart, farming is all about community, and this book is a way that we can give back to that community, and help it grow. For anyone looking to try their hand at sustainable, renewable farming practices for non-industrial farms, this book is for you.

We know firsthand the immense value of raising poultry for a family, including how it can help make ends meet and provide healthy food for the community. We also know some specific challenges and pitfalls that need to be overcome to succeed. Sharon and I hope to offer you and your family the necessary tools to give yourselves the best chance at this rewarding endeavor – and highlight what not to do.

When we started working on this book, we decided that you should know our history, because you should always know where your information comes from. Some of the things we went through can be replicated, others are unique to us. But with enough info, you can at least see where we're coming from.

The journey to our family farm started with my grandparents in the 1940s. They left Spokane and settled in the Dungeness Valley in Washington state, located on the Olympic Peninsula. There, on some truly picturesque land just across the water from Victoria, Canada, they started a successful dairy farm. My mother followed in their footsteps, and she passed down her love of animals, open air, and farm living to me.

I spent my summers there, playing in barns and learning all about farm life. The memory of fresh salty air blowing in from the Salish Sea, mixed with the scent of fields and animals, left an indelible mark on my childhood. It was a Norman Rockwell painting come to life.

Despite my love for it, when I grew up, I did not continue on the farm. In college I went in a much different direction, and I started a career in broadcast television and camera work as an intern in southern Oregon. From there, I moved to Portland and built a career working on action sports and reality shows like "Ax Men" and "Ice Road Truckers." I had some fantastic adventures while pursuing this career, but the rewards were always a mixed blessing, as it meant I was forced to be away from my family for extended periods of time. We realized that our family needed a better, more rural life, and we wanted our boys to know where food came from and how to raise it. So in 2013, we decided to change everything.

We fully appreciate that not everyone can do what we did. We were fortunate to have a great support system and experience to draw on when we decided to jump off the proverbial cliff, and the circumstances worked in our favor. But we also know that sometimes when something calls to you, you must answer the call or forever wonder what might have been.

With our path decided, we sold our house and temporarily moved to my family's farm. We went deep into research, and embraced solid, real-world work. We had begun our "farm college" in earnest.

Our goal was to learn how to become sustainable farmers before we purchased a property of our own, and working on the family farm allowed us to get ourselves ready. That time was critical in many aspects.

Farming can be backbreaking work, sometimes literally. A few months into our endeavor, I was logging the property with some llamas and lost awareness. I was taking my 30-something-year-old body to the brink, over and over, until one day I bent over to grab my chainsaw and tweaked my back. That should have been the time to take a break, but I had so much more to do, so I continued for two more weeks. If you've ever worked on a farm, you know that things don't stop, but I should have done things differently. That lesson cost us dearly.

I worked myself until I was crippled. And since I didn't have insurance, back treatments quickly burnt up the nest egg we had saved from selling our home. We went from the farm dream to a nightmare in a moment. It was a setback that seemed catastrophic

I quickly fell into a terrible depression. The injury prevented me from working on or off the farm and I started having ugly fantasies of Great Depression-era famine and destitution as the farm threatened to die around me. It was a dark time for me and things looked bleak. Then, one day a friend came to me and said, "Hey, I am happy to see you." Sometimes, the greatest gift you can give someone is to tell them you are happy to see them. I found hope again. I owe it all to friends and family, and it wouldn't be the last time.

I went through months of debilitating pain. More than once my screaming woke Sharon in the middle of the night. Time and again she had to take on multiple duties, including nurse, as she pulled and extended my leg to relieve the pressure on my back. Thankfully, I finally found a doctor who knew how to help.

I clawed myself out of the dark hole I was in, landed some work, and we found a way through. It sidetracked the full-time farm dream for a few years, but it didn't crush our spirit. After all, nothing is better than open country air, a rooster's crow, and frogs singing you to sleep at night to let you know that you've found true paradise. We just adjusted our goals and turned farming into a hobby while we slowly worked toward our dream again.

Eventually, our journey took us away from the family farm on the Olympic Peninsula and closer to Portland, where we could be near our family and friends. After searching high and low, hoping to find a property we could afford, we ended up at a five-acre horse pasture in southern Washington. It took time and effort, but we eventually turned it into a "market" farm. It is the perfect place in the country and closer to our family and support network.

Who We Are

We believe in small-farm agriculture, and the importance of purchasing locally-sourced food and farm products. We know that's sometimes easier said than done, but the more support we can show our community, the more options there are for everyone. With that in mind, we have tried to do right by our customers, while nurturing the local environment. Our goal is to provide high-quality, pasture-raised, humanely treated meats, eggs, and produce, all raised on our five-acre farm in Battle Ground, Washington.

Our land was not initially created to be a farm. Ideally, a functioning farm should be a single ecosystem, with each section helping to sustain the next, and converting it has been physically and financially daunting. Sharon and I have built it up step-by-step over the years, and thanks to new structures and old tractors, we finally began to hit our stride.

We didn't actually set out to become a market farm. We originally just wanted to live and sustain ourselves off the land, but the COVID-19 pandemic and lockdown encouraged us to focus on our community. We left our jobs and became full-time farmers, but this time we entered with a fair amount of caution and a lot more knowledge from our years of hobby farming. The question then became could we produce enough for others? Could we pay the bills and feed ourselves? I'm happy to say the answer was "yes," and so can you.

We have a vision of small farms run by independent farmers, which is a big part of why we wrote this book. We support small, independent farms that provide food locally, but think globally. Both personally and professionally, we support regenerative agriculture producers, those who treat their animals humanely, and strive for low carbon footprints. We also want to support our neighbors, local food networks, and the regional farming economies. By supporting and practicing eco-conscious farming, we can help combat global warming and food insecurity, even if it is only to a small degree.

Being a small, independent farmer means you are part of a community, but when it comes to the day-to-day work, you'll probably be working in small groups, or even solo. That means you'll need to work smarter, and find the most efficient ways to handle your workload. One of the main goals of this book is to offer a blueprint of sorts that other small farmers, and those aspiring to one day become a small farmer, can use. Small acreage farmers have the ability to do a lot with their land, and raising 1,000 chickens for their local markets is both realistic and sustainable.

That leads us to this book. It took several years (and lots of mistakes) to create and refine the Misty Frog Acres method for raising 1,000 chickens in one summer with just two people, but hopefully it can help others. A college professor once told me, "You can do anything. You just have to pay the consequences." The consequences of not following your dreams far outweigh the consequences of doing them.

1

Land and Infrastructure

Before we get into how to raise 1,000 birds in one summer for farmers markets, there is a great deal to review. Maybe you already have a system in place, and you are just looking for some pointers. Maybe you are new to raising meat birds, or even birds in general! This guide is designed to be easy to navigate regardless of your experience and knowledge base. If you just want to jump into the how we did it, look for the "Danny's Details" in each section that clarify what we did.

Before you commit to acquiring land, you'll want to plan what to do with that property. You'll need to understand how much area you will need, the work involved to prepare the land, and how to make the land work efficiently for you. Does the land have the required infrastructure, or will you need to build that out? Is there sufficient water access? What is the erosion and drainage profile like? How close is the property to things like hospitals, veterinarians, and schools? All these questions must be asked and answered before you set your family on this path.

RENTING VS. BUYING

Danny's Details: Our adventure in obtaining property consisted of a little bit of luck in the family, getting our elbows and knees dirty, and a little creative thinking. When we first started looking for our current property, we were driven by the need to be close to loved ones for health and support reasons, and ensuring we had reasonable proximity to non-farm work. After my back injury and subsequent draining of our nest egg, we searched for a property near Portland, Oregon, a city we both had deep ties to. This was where the luck and elbow grease came in.

Sharon's cousin happened to be looking for a renter on her five-acre property. It had a bunch of weed-covered pastures, countless blackberry bushes, an electric horse fence, a broken-down barn, and a single water spigot. After we moved in, we discovered that her cousin wanted to sell it, but couldn't. There was an over-aged manufactured home on the property, and substantial code repairs were required to make it acceptable to a loan servicer. We agreed that if I did the repairs while we rented, she would sell to us at what was left of the loan she owed.

I reroofed the decrepit garage, installed earthquake strapping to the foundation, and did other repairs that were required. It was a blessing that Sharon's cousin was willing, able, and kind enough to help us during our dire straits, and we were able to help her back. Farming is so much about helping each other and your community. Don't be afraid to look for situations where you can leverage your skills, family, and friends for an endeavor like this. Don't forget to always be thankful, and always pay it forward.

Buying a property will allow for you to efficiently plan for the long-term, including upgrades. As you learn, you grow; with that growth, your infrastructure requirements will naturally mature. Plus, owning the land gives you a sense of pride and connection that is hard to replicate on rented property, and it's also something you can pass down to another generation. For those with children, it's a powerful incentive.

Sure, not everyone can own land right away, and renting is a viable option if you lack the up-front money. But most farms will require you to make significant changes to the land. Owning it obviously bypasses that issue, but if that isn't an option, you may want to try to negotiate a special deal with the owner that allows you the freedom to make changes as needed without constantly seeking permissions.

Renting can also make sense if you are new and uncertain about your final end-goal farm. If you have any uncertainty in this respect, renting can let you test out the space you need, the farming methods to use, and even what kind of plants or livestock you want to deal with. This can provide you with the necessary experience to take you into that second or third year, but be aware that you will be sacrificing the infrastructure and development when you leave. But if you are unsure, this is a great way to start small.

ZONING AND REGULATIONS

Whether you rent or buy, you'll need to know if the land is correctly zoned and within regulations to support your needs. What kind of farming activity, buildings, improvements, and infrastructure can you put on your land? What kind, and how many animals of any specific type can you handle? Do your research. Every country, state, county, or province can have different stumbling blocks.

For example, within the city limits of Seattle, only eight fowl are allowed on residential properties. But in King County where Seattle is located, outside of the city limits it depends on the size of the property. There may be laws related to what you can and can't build on your property, what needs to be licensed (sometimes that's broken down by size so that small buildings do not require licensing but larger buildings do), water rights, land uses, and many other issues. If you are unaware, these can cost you time and money, so educate yourself.

UNDERSTANDING COSTS

For all intents and purposes, farming is a small business, and like most small businesses, don't expect to make money your first year, or even the second. There are exceptions to that rule, but make sure you have enough savings for up to two years, or barring that, plan on keeping your day job to maintain a steady flow of income.

My situation allowed for me to primarily work on the farm during the warm seasons, then during the cold seasons I would take jobs as a handyman, a substitute teacher, and also take on more freelance work in broadcasting. I did this until we were ready to go "all in" on farming. There is nothing wrong with slowly easing out of your "regular" job and moving to farm work, as you get your feet under you and finances balance out.

To put it mildly, understanding the balance between costs and income is a critical factor before diving into any endeavor. We will discuss the actual sales and profit model in "Chapter Eight, The Business of Selling," but for now, we need to talk about costs concerning land.

BASIC INFRASTRUCTURE

Danny's Details: When we started on the Battle Ground, Washington property, it was a weedy pasture with lots of random blackberry clusters and weird, buried garbage. At this point we had a few years of experience on my family's well maintained, albeit unused, generational farm in Sequim, Washington, so we knew this was going to be a serious undertaking.

In Sequim, the property was sort of frozen in time from when it was a farm previously a farm. We had shops, chicken coops, milking infrastructure, tool sheds, hay barns, loafing sheds, irrigation, and other structures that in some cases have stood since the 1890s. On top of that, I had plenty of examples of "how Grandpa did it." Compared to the level of infrastructure in Sequim, what we were moving to in Battle Ground was a drastic downsizing.

Our new barn (new to us, that is) was at least standing and had electricity, though it needed repair. There was one water spigot. There were some fence posts and fencing, but we needed a lot more to set up our pastures for rotational grazing and separating ruminants from poultry. Fortunately, the main barn infrastructure covered the basics we needed to keep the animals alive. It had water, power for brooders, and a dry, non-drafty shelter. Other than that, there wasn't a lot. In time, I built more covered areas, a shop, loafing sheds, chicken tractors, and other infrastructure. It was a lot of work for us, and it's a great example of what to be aware of when looking for a property. Keep an eye on how much work it will take to get a property up to where you need it to be.

Unless you are fortunate and can purchase or inherit a successful generational farm in the same area you wish to work, you will likely have little or no infrastructure. For the purposes of this book, infrastructure includes barns, shops, brooders, chicken coops, chicken tractors, mowed or grazed pasture land, fencing, electrical access, and water access. Your own needs may vary, but those are the basics. Depending on the property, there may even be things like access roads or wells that need to be installed. Don't forget the soft costs like licensing, permits, or inspection fees either. Go into any property with your eyes wide open, and don't assume that an amenity readily available in a city suburb will be available on an agriculture property.

Most likely, you won't be able to pay for every improvement from the get-go. Carefully evaluate any existing buildings or infrastructure already available on the property. Brooders can go into garages until a barn is built later. Water buffaloes and rain catchers can serve while you wait to install ground piping or an additional well. Also, consider carefully if you can do a project yourself since that will save you money and allow you more flexibility in getting your farm life underway.

Ultimately, though, there will be things you require before you get your first bird. Do not have your chicks show up until you have somewhere to put them, a way to keep them warm, and food and water available. It is your responsibility to take care of these little lives, and since they are making the ultimate sacrifice for your family, they deserve the proper care.

LAYOUT

Danny's Details: Layout is important, but each property provides different challenges and benefits. We were lucky with our property layout. The land was dominated by the existing pastures' long, rectangular north-to-south shape. That let us break the five acres into different paddocks to rotate between where the meat chickens and sheep were pasturing. We kept the sheep six rotations away from the chickens, with the chickens always following the sheep. This reduces the possible bugs, bacteria, and viruses that poultry can inevitably pass to other creatures if proper biosecurity isn't practiced.

Here you can see five different areas: the furthest is the finishing tractors' paddock; to the left are the low pasture tractors; on the right are two paddocks that are resting.

Another benefit of our layout is that our house is located on a hilltop on the south end, overlooking the rest of the property. This gave our dogs (we have a rough coated collie and a border collie) an advantage to easily see if something was out of place and raise the alarm when necessary, or head off to investigate. Between the two, they scared off many raccoons, coyotes, hawks, and even owls with their teamwork. The dogs provide great security for our farm, and it is really fun to watch them work.

The final lucky stroke was that the barn is positioned about halfway down the property. Being located at the center of the property gives us more options on how to move our chicken tractors, address animal rotation, and generally handle our animals. It is also where the main spigot is, making it relatively easy to get water to most places with hoses.

I encourage you to lay out your field map, garden map, and structures, then focus on long-term installations. Spend the time and money on those first. I built WAY more salvaged-material chicken coops than I ever needed, and bought way more hoses than I should have, all to try and save a buck. Rather than becoming an operational benefit, it turned things into more of an expensive shanty town. While construction salvage and building something from stuff you find is excellent, make sure you spend the proper time and money on things that need to last.

Your fields will inevitably become a critical piece of the puzzle. By the time the chickens begin to use them, you will have already spent a lot of time and money, so make sure that they are ready. Are food and water within easy transport? Are there roads or paths that need to be constructed? Hauling multiple five-gallon buckets of water two hundred feet from a hose bib is great for getting in shape, and also throwing out your back. Be smart and make sure the work is done in advance.

PASTURE

Though there are other ways to raise broiler chickens, we advocate using pasture methods. Raising chickens on pasture will encourage a healthier meat product, which is one of the major draws over factory-produced poultry. Be sure to have a good mix of plants within your pasture, including legumes, broadleaf plants, grasses, and grains.

Matching plants to soils is essential and will differ on almost every farm. You should consider the slope, internal drainage, water supply, nutrients, and soil depth. Choose species of plants with long growing seasons, are resistant to grazing (specifically they have the ability to recover after grazing and trampling), and are perennial (meaning they come back year after year without replanting). The following are the four types of plants that will provide a good mix for your poultry:

- **Legumes** might include plants such as true clovers, medics, sainfoin, bird's-foot trefoil, and vetches. White clover should probably be one of the favored plants for poultry in your pasture.

- **Broadleaf plants** are known as "weed" species, such as plantain, dandelion, and dock. Be aware that some broadleaf plants can be toxic, so research the common types in your area and know how they will affect the various animals that use the pasture.

- **Grasses** are another important plant for most pastures, though they will be more beneficial for ruminant animals (cows, sheep, and others that can digest grasses). Chickens do not fall within this category, but the grasses will provide excellent cover for insects like grasshoppers for them to eat. Grasses can also be excellent ways to combat erosion, and resist denuding an area.

- **Grains** are most beneficial in early spring, especially for farmers who use chickens in rotation with row crops. Chickens are great at tilling the soil, and provide a natural fertilizer. The best grains to consider will depend on the land and climate, but oats and rye are typically good choices.

Keep the height of the pasture short, no more than a few inches tall, so your poultry have an easier time foraging, and the chicken tractors have an easier time navigating the terrain. Mowing or rotating other livestock through the pasture is the best way. The poultry will also help to limit flies and other pests that would negatively affect your other livestock. Always use your animals and plants in concert with each other to gain more benefits.

Research how to rotate your pastures with the animals you have. For instance, on our farm, we keep sheep in front of the chickens. This is because if you switch the order and put the chicken in front of the sheep, the chickens may leave behind bugs that can make the sheep very sick. One year, we lost five sheep to a nasty round of coccidiosis, probably because the animals cross contaminated fields. We also wait about 50-60 days before our sheep would be on a paddock the chickens had been on previously.

Presumably, most farmers will use an existing pasture area and won't be building from scratch. If you fall into that group, you can incorporate the above suggestions over time, as you find the resources. If you find yourself in a position where you're clearing or expanding new land, you'll have to put in more time and work, but it will give you an opportunity to establish your ideal pasture from scratch.

If you are clearing land, you are most likely dealing with underbrush and trees. Harvest the timber for your use or sale, which can help offset costs. Take the time to check the land carefully for rocks that may cause damage to any equipment you use in the future. If you can get your hands on some goats (many areas offer them to rent), an attractive option is to cordon off an area and let them graze the underbrush before tree harvest. Remove stumps, then establish the next cover (pasture). You will likely have to till the soil and add nutrients/fertilizer before planting your new pasture.

WATER

Evaluating and planning the water situation and availability on your land is paramount. This includes the amounts, the quality, and the source of the water. Do a little research and try to understand the drought history of the land and area. Consider storing water in on-site ponds or tanks during high-water availability.

Water quality analysis of wells (sometimes required by health authorities), runoff, standing water, and other water sources can be important to consider. You should test for the following before use in poultry production:

- **Hardness**: Calcium and magnesium salts can cause water to become hard. This condition may cause scaling and other buildups that could clog water lines. If water is high in sodium, avoid water softener, unless potassium chloride is used instead of sodium chloride.

- **Iron (Fe) and Manganese (Mg):** While iron and manganese won't affect poultry health, high levels can result in line build-up, leaky water nipples, and stop-up foggers. Treat iron with the addition of chlorine, chlorine dioxide, or ozone before mechanical filtration. Manganese is similarly treated, but has a slower reaction time and requires longer contact with chlorine before filtration.

- **Nitrate (N):** The presence of nitrates in water could result from decaying organic matter and should be an indicator to check for bacteria. To treat, add chlorine, chlorine dioxide, or ozone before filtering.

- **pH:** The pH measures acidity, and scales from 0 to 14. A value of 7.0 is neutral. Below 7.0 is acidic, and above is basic. 6.0 to 6.8 is ideal for broiler production, but birds can tolerate a range from 4-to-8. Anything higher can result in lower water consumption. Raise pH with soda ash, lime, or sodium hydroxide. Lower pH with phosphoric acid, citric acid, or vinegar. Low pH can also be reduced by removing free carbon dioxide through aeration.

- **Alkalinity:** This condition results from calcium carbonate, bicarbonate, or sulfate. Treat with acidification or an ion exchange de-alkalizer.

- **Toxic Compounds:** Elements such as lead, selenium, and arsenic should be kept below 1.0 ppm to prevent poultry health problems and residues. Treat by checking the water system for old pipes, fittings, or other methods of contamination and remove them.

- **Bacteria:** Some bacteria are inevitable and acceptable, but it should be kept at a minimum. Coliform bacteria in drinking water can be an indicator of fecal contamination. Iron bacteria is not a health hazard but can cause a reddish-brown slime that coats pipes, affects pumps, and plugs drinkers. Clean the system with sanitizing cleaners to treat and establish a daily water sanitation system.

Water quality can alter during periods of heavy rains and drought. Re-testing can be needed if you begin to notice a change in the health of your birds. Change filters and consider water treatments (for water with high iron) where filters may not be effective.

Your water source can come from several different sources, and each should be considered on their own merits, as opposed to what is required to raise your poultry. Most chickens require about one pint of water per day, more if it is hot. In a later chapter we will discuss water directly in relation to your chickens. It's important to know if your land, or the land you are considering, has sufficient water flow for all of your farming needs.

- Drilled wells are a clean water source. Barring unexpected contaminants, most municipalities carefully track aquifer water, depth, and yield. Be aware that the location, depth, and water quality can vary, even on the same site, and may be affected by the season and heavy use.
  Most well drillers must file reports on the wells they work on with local authorities, so records are available.

- Surface water includes streams, rivers, lakes, ponds, and drainage ponds, which depend on local land or spring runoff. These sources can vary depending on rainfall and season. Surface water sources may experience a higher level of contamination with road salt, industrial or agricultural chemicals, algae, and plant pathogens.

- Rainwater can be accessed with cisterns or tanks that collect runoff from roofs or rain catchers. It is generally clean, barring debris getting into the system or local air pollution.

- Municipal water comes from the city, country, or municipality. This water is used for residential drinking, resulting in high quality and cost. The main concerns are how much you'll pay, the types of treatments used, and whether the supply is guaranteed during drought.

Danny's Details: When we started our farm, we only had access to a single water spigot. I made water collection tanks out of any clean water container I could get my hands on, and deployed them as needed. Sometimes I used a permanent 50-gallon barrels hooked to roof gutters on the barns, other times it would be in a small 25-gallon tank with a spigot attached to the ATV or EV tractor. Other times, it was an entire 250-gallon trailer in a "water buffalo" style setup (more on that later), with irrigation pumps and power that I could park in the lower field.

Every collection tank had its uses, but it was always about trying to get enough water in a position where I could minimize hand-walking water to the chicken tractors. The further away from the clean water sources, the larger my water tanks would be, because meat birds need about 25 gallons per grouping of one-hundred (about four chicken tractors worth). After several years of doing it, the best setup I found was a permanent 250-gallon tank at the furthest chicken coop, and a 250-gallon water buffalo with a pump and hose hookup. This allowed for about ten days of water in the furthest field, ten days in the middle, and the rest would be close enough to the main barn to provide for the babies and pullets that hadn't started their journey on the pasture.

For the sake of your poultry, you should address standing water on the property. Many viruses and bacteria love damp, moist conditions, so if there is a bad year for avian bird flu or some other bug, a dangerous vector is standing water caused by compacted soil. Avoid those spots with your chickens when water is present, or if the soil moisture is high. These conditions most often occur early in the spring pasturing season. Combined with temperature fluctuations, this can lead to the loss of birds.

A good way to mitigate compacted soil and allow moisture penetration is to find those spots and then spend some time aerating, dethatching, or breaking the ground deep with a plow. You can also dig small drainage ditches out of the field into the main ditch area, or downhill from the affected areas. Some grasses and plants like water more than others, so adding these can help get surface water off the upper layers of the pasture. Mineral additives like lime and other groundbreaking products can help as well. Eventually, a well-rotated pasture and the inclusion of organic matter – i.e., poop – will help loosen hard-packed soil, but those results take much longer to realize.

Regarding layout, other questions you should ask are: How fast are the birds soiling the footprint of their pens? How often do I need to move it to keep them healthy, and on fresh grass? How long does it take to grow back fresh grass on a soiled section? Should I use a tractor or ATV to haul stuff to them when they are further out? Can I mitigate the risk of predators by moving the pens a certain way, or putting them in a circle around a safer area? I will try to help you answer these questions, but consider your specific land and mobility options while you start laying things out.

SKILLS YOU NEED TO KNOW

We all come from different backgrounds. Some of those reading this book will be steeped in farming knowledge, and others will be fresh from the city with baby-soft hands. Either group has the potential to successfully run a farm, but one will have a much better chance at success than the other.

It is still possible to succeed even if you don't have all the required skills when you start planning your farm, but you WILL want to acquire them before actually diving in.

- **Organizational Skills:** Farming is like any other business, requiring an understanding of various dynamics. Cash flow, paperwork, machines, maintenance, daily or monthly routines, and other activities require organization. Keep organized records of contracts, insurance, certifications, labor, repairs, machinery serviced, costs, and income. As a small farmer, you will most likely do this all yourself, but if you hire this out, go through reputable agencies.

- **Farm Operation and Management:** Farming consists of daily, weekly, monthly, and yearly routines. When do you order your chicks? How much feed and medicine do you need for the chickens you want to raise? How do you treat your poultry during cooler or hotter months? Do you know how to handle disease management for your animals? When should you do things yourself, and when should you use a professional? You need to understand what is required to run everything smoothly.

- **Technical and Mechanical:** This skill is crucial, especially for small farmers. You will not be able to afford specialists to handle things like IT (information technology), drivers, mechanics, and many other roles. If something breaks down, or you need a solution that is not readily available, you must fix it or jury-rig it. Do you know how to order chicks or feed online? Get those skills to help protect your profits in the long run.

- **Interpersonal and Communication:** While the image of the grumpy farmer with a shotgun is common, that should not be your reality. Know how to talk to, and relate with everyone involved in your business. Examples may include suppliers, customers (important!), employees, other farmers, and partners. If you have friends, neighbors, and family who want to offer a hand, I encourage you to practice your interpersonal skills.

- **Marketing:** Good products do not automatically sell themselves. Therefore, you must know how to create awareness in the marketplace for your product. For poultry, do you know the benefits of selling whole birds vs. parted birds? What farmers markets are in the area, and how do you establish yourself? Is there a market for discarded parts (like pet food makers) that could buy your product? Knowing how to access these markets will be important for your small farm.

- **Always Learn:** Be open to learning new or more effective techniques in the areas of expertise listed above, along with others not listed. New technology, techniques, and tools emerge every day. Products meant for something else may prove useful to something you are working on, with a little finagling. For example, something like drip tubing and timers for watering gardens may prove useful for hooking up to your chicken tractors. If you don't take advantage of them, your competitors may leave you behind.

Danny's Details: When we first became interested in chickens, Sharon and I lived in the city. We had three chickens, which quickly became ten chickens. Chickens are the gateway farm animal! From there, we became the neighborhood experts. We researched and handled animals constantly. We went to the local feed stores and befriended the owners and workers. Sharon eventually picked up a job at the Urban Farm Store in Portland, where she gained more experience and knowledge with feed, along with other critical items that would come into play when we were finally in the farming business.

Much of my understanding of farming came from talking with my stepdad, a turkey farmer in his early years, along with my uncle and mom. They all had fond memories of farming from when they were kids, and did me the great honor of passing their experience along to me. Once we got to Farm College (what I call our family property up in Sequim), the learning was immersive and hands-on. We met other young local farmers and shared in their experiences. We were able to use tractors for the first time, and more importantly we learned to fix them. If a weld broke, I had to learn how to fix it on my own. If you don't know much about pumps, water rights, or plumbing, you'll learn fast when you go down this road.

Once we moved into the property in SW Washington, I needed to learn new skills on my own, without the help of my uncle and his knowledge of equipment. One day, he showed up on our farm with an old tractor and various tractor implements. I had never dealt with older tractors before, so I asked him about it. All he said was "I don't know anything about green tractors, good luck." Then he handed me an old manual and the keys to the tractor, and drove away.

In retrospect, I think he was deliberately forcing me to develop some mechanical skills. I had to learn everything about it from the ground up, and it ended up being my must-have piece of equipment. I learned how to replace the clutch by getting on a tractor enthusiast's forum and asking for help. I also found a new frame there when the tractor broke in half. I learned how to become a tractor mechanic from various internet groups and YouTube channels. That tractor is still in use on our farm. Its twin is also there, which I hand-converted to an all-electric vehicle. There is a vast amount of knowledge out there and folks who want to help, you just have to put on your farmer cap and go find them!

2

Choosing Your Chicken Breed

Now that you have your land, and hopefully a good start on the infrastructure to support some chickens, you must choose a bird breed that will work best for you. There are many breeds to choose from, but depending on your methods and ability, some breeds may be superior to others.

When you're choosing a breed of chicken, there are many questions. Are the birds heritage or commercial? What is their temperament? What is their growth rate? How do they do in cold or warm weather? Am I raising them with a pasture, enclosure, or free range? These questions are important, and can mean the difference between failure or successfully bringing the birds to market at a profit.

The term "heritage" describes non-commercial breeds in danger of extinction. To be considered a heritage breed, chickens must meet American Poultry Association (APA) Standard Breed guidelines (the genetic line is traced through multiple generations), be naturally mating, have a long and productive outdoor lifespan (five years for breeding hens and three years for roosters), and a slow growth rate of no less than 16 weeks to maturity.

Heritage breed chickens originated from the needs of small farms, over hundreds of years. They are hardy (higher disease resistance and vitality), adaptable, able to fend for themselves, and capable of withstanding fluctuating temperature differences.

While these birds can be a good fit for your farm for general egg laying and small-level meat production, market meat production may prove difficult, resulting in higher costs to raise these birds to maturity and sale.

Hybrid breed chickens are crossbred from family lines with particularly desirable commercial-friendly features. Some may have been developed for egg-laying capability, while others, called "broilers," are made for meat production. It is these latter types of chickens that we want focus on for this book, as their primary role is to help you bring food to your community.

Danny's Details: When it comes to raising chickens, there are several varieties you may have an opportunity to raise. When we first started raising meat birds, we went right to Cornish Cross as the chicken of choice. No other breed grows as quickly, and they can reach butchering weight in eight weeks. That's a key element for our "1,000 birds in a summer method." No other breed provides so much weight for the cost of raising them. We consider most other breeds as niche, so we went right to the king of meat birds.

In addition to Cornish Cross, we also raised quails and turkeys during our first year. Quails are quick to reach butchering age at six weeks; turkeys take fourteen to eighteen weeks. Turkeys also tend to be very tame and easy to handle. They are more delicate than chickens when babies, but they are quite hardy once they are feathered out.

Cecile Long Steele of Sussex County, Delaware, is often cited as the pioneer of the commercial broiler industry. In 1923, she ordered 50 chicks for egg production, but received 500 instead. She kept those chicks and raised them for meat. Her small business was so profitable that, by 1926, Steele built a broiler house with a capacity of 10,000 birds. In 1952, broiler chickens surpassed farm chickens as the number one source of chicken meat in the United States.

Tristan Lulay

Cornish Cross (Cornish Rock):

Age at Maturity: *8–10 weeks*
Avg. Mature Weight: *9–12 lbs (with about a 5–10 lb processed weight)*

Cornish Cross chickens are a hybrid of the Cornish and White Plymouth Rock breeds. They are one of the most successful and widespread meat birds in production today, due to their good dispositions, rapid growth, and the high amount of meat per bird. If you buy chicken in the store, it most likely comes from the Cornish Cross breed. They are ready for harvest between 8-10 weeks of age. On small farms, Cornish Cross birds are best raised on pasture within an enclosure, as they cannot move quickly enough to evade predation.

One concern with the Cornish Cross breed is that they are susceptible to health problems, like heart and leg issues, due to being bred for very rapid growth. Most Cornish Cross will die of heart issues if they are allowed to mature past 8-10 weeks. They also do not tolerate hot or cold temperatures, or do well at elevations above 5,000 feet. There is a common concern that Cornish Cross have trouble self-regulating how much they eat, but I didn't find this to be a constant problem. Only in the last two weeks of their growth cycle did I begin to regulate the food, giving them enough so they would run out by nightfall.

Cornish Cross are not the best choice to raise for farm breeding. They are generally too large by the time they reach sexual maturity, and their eggs will most often be infertile. If you choose Cornish Cross, you must almost exclusively order chicks from the hatchery.

Red Ranger/Freedom Ranger/Rainbow Ranger/Big Red Broilers

Age at Maturity: 12 weeks
Avg. Mature Weight: 7-10 lbs (with about a 3-8 lb processed weight, depending on the bird's gender)

The Red Broiler birds are highly popular, but the names can be confusing. They are a conglomeration of breeds, and their specific genetics are somewhat murky. The name differences come from various hatcheries trying to differentiate themselves from others, creating different name brands. These hybrid birds are good for meat production, but can also become egg layers. These are popular meat birds, and can be great alternatives to the Cornish Cross. Rangers are good foragers, so depending on how they are raised and fed, they may require less cost to feed. They do, however, take 12 weeks to age enough to harvest. Due to their genetics and slower growth period, Rangers tend to have fewer health problems (although they still have a few). Rangers do well in most climate conditions.

Red Rangers are another bird unsuitable for breeding, as most of their eggs will be infertile. Also, the hens are much smaller than the roosters, which can make a difference in processing time.

Bresse

Age at Maturity: 16 weeks
Avg. Mature Weight: 5-7 lbs (with about a 3-6 lb processed weight)

Bresse birds are a heritage chicken originally found in the Bresse region of France, though you can buy them from U.S. hatcheries. Though they are a more expensive breed, they are prized for their marbled and high-quality meat. Considered one of the best-tasting breeds, this bird has thin bones, resulting in a higher meat-to-bone ratio. Their temperament is considered non-aggressive.

Bresse chickens are excellent at foraging for insects and grass seeds. They can find their own food, but appreciate a daily ration as a supplement. They are a fairly hands-off breed, but do need protection from possible predators.

Bresses tend to take about 16 weeks to reach harvest maturity, and to build the traditional Bresse flavor for the meat takes a special diet.

Turken

Age at Maturity: 20 weeks
Avg. Mature Weight: 6-8 lbs

Turken, despite its name, is a pure chicken. They are also known as Naked Neck chickens due to their naked heads and necks, a trait bred into them to make them easier to pluck. These meat chickens do well in cold climates (despite their lack of feathers), but if temperatures drop below freezing you must watch their combs as they can be susceptible to frostbite.

Turken mature to harvesting age at about 16-20 weeks, which is on the longer end for a meat bird, but they are good foragers and tend to be docile. They are good egg layers if they are allowed to reach maturity.

Fresh Nest Farms

Ginger Broiler

Age at Maturity: *12 weeks*
Avg. Mature Weight: *8–10 lbs*

Ginger Broilers are a heritage bird primarily raised for meat. They grow quickly and have fewer health issues than the ever-popular Cornish Cross. They are a good bird for beginners to raise, as they are friendly, hardy, resistant to disease, and do not require much special attention. Their meat is not considered overly flavorful, but they are still desirable meat chickens.

Jersey Giant

Age at Maturity: *20 weeks*
Avg. Mature Weight: *10–13 lbs*

The Jersey Giant was an attempt to replace turkeys. They are a heritage breed that can be slower to mature than other heavy breeds, but are also good for egg-laying. Their large size makes them a good choice for commercial and small family farms. This breed is known for its good foraging skills and can be an excellent free-range bird, while being calm and friendly toward humans. In general, Jersey Giants are healthy and hardy.

Fresh Nest Farms

New Hampshire Red

Age at Maturity: *16–20 weeks*
Avg. Mature Weight: *6–8 lbs*

New Hampshire Reds are a dual-purpose breed derived from the Rhode Island Red and Plymouth Rock Breeds. They are a popular choice for meat production due to their relatively high growth rate and solid harvest weight. They do well on pasture, confinement, or free-ranging, and can withstand both warm and cold climates. As a breed, they tend to be more independent and prefer to be left alone. They're also known for being broody and quiet, and males can be aggressive if they feel threatened.

Plymouth Rock

Age at Maturity: 20-24 weeks
Avg. Mature Weight: 6-8 lbs

The Plymouth Rock breed is considered one of the most popular and versatile breeds worldwide. It was bred for the mass production of meat and eggs, is fairly low maintenance, and it can survive cold climates. Unsurprisingly, it's one of the two primary breeds used to create the Cornish Cross. While their temperament is generally docile, the roosters can be aggressive with other chickens. Overall, this makes them good beginner-level birds.

Fresh Nest Farms

Buckeye

Age at Maturity: 16–21 weeks
Avg. Mature Weight: 6–9 lbs

Buckeyes are a hardy breed of chicken that can live in cold climates. They are fairly healthy, and can resist most diseases. They are also excellent foragers and their temperament is friendly, which makes them popular chickens for backyard farmers. Buckeyes are dual-purpose chickens, so they lay eggs and are harvested for meat. Although they can take some time to reach maturation, it's worth the wait. Buckeye meat has been described as "nutty" and tastes exceptional when brined.

Melinda Sayler, CC BY-SA 3.0 <http:// creativecommons.org/licenses/by-sa/3.0/>, via Wikimedia Commons

Chantecler

Age at Maturity: 16 weeks
Avg. Mature Weight: 7–9 lbs

The Chantecler is a dual-purpose chicken from Canada. These birds tend to have a gentle and calm nature, while also being able to withstand cold climates. They are popular for their meat because they mature faster than other breeds, and they are great at foraging. For small farmers, they are ideal, because they don't require too much attention.

Chelsea Beck

Langshan

Age at Maturity: 16–21 weeks
Avg. Mature Weight: 6–9 lbs

This bird is a heritage breed that originated in Asia. They have the potential to be a dual-purpose breed, but they are primarily known as meat birds. Due to their size, they don't do well in very hot conditions. They also do better in dry, sheltered areas rather than free-range. Langshan are quite docile, calm, and friendly.

Matthew Vearing, CC0, via Wikimedia Commons

Delaware

Age at Maturity: 16 weeks
Avg. Mature Weight: 7–9 lbs

Delawares are resilient birds that reach maturity swiftly. Unlike prevalent commercial meat breeds, Delawares do well in free-range environments. They are generally of a calm temperament.

Josh Larios from Seattle, US, CC BY-SA 2.0
<https://creativecommons.org/licenses/by-sa/2.0>,
via Wikimedia Commons

Orpington

Age at Maturity: 20-22 weeks
Avg. Mature Weight: 8-10 lbs

The Orpington is a popular, dual-purpose heritage breed that does well in cold weather, but may need help in warmer climates due to its size. They tend to grow fairly slowly and have a gentle, calm demeanor without any aversion to humans. They are also good foragers, and do not generally require a high-protein diet.

Lavender Orpington pictured

Fresh Nest Farms

Brahma

Age at Maturity: 16–20 weeks
Avg. Mature Weight: 9–12 lb

The Brahma breed of chicken is a large bird that can also successfully be used for egg laying (able to lay in winter, too). Their size makes them ideal for colder climates, but presents challenges in hot weather. Low fencing is sufficient to keep this bird contained, as they are not known for flying. They are solid foragers and can free range, but they can also be kept in confinement as long as their shelters accommodate their larger size. Their temperament is generally docile and easy to take care of.

Bodlina, CC BY-SA 3.0 <https://creativecommons. org/licenses/by-sa/3.0>, via Wikimedia Commons

Leghorn

Age at Maturity: 16–21 weeks
Avg. Mature Weight: 6–8 lbs

Leghorn chickens are an excellent dual-purpose beginner-level bird that does well in both confinement and free range. They are great foragers that do not require much feed, especially if allowed to be free-range. If used for eggs, their feed-to-egg ratio is well regarded. Leghorns are active birds that tend to be flighty and nervous, but they are fairly docile.

3

Brooders and Receiving Your Chickens

So now you have a box of chicks! Or do you? Before you start raising your chickens, you need to decide if you are going to hatch them yourself or order them from a hatchery. We used Cornish Cross for meat birds, which are almost impossible to breed outside of a specific environment, so we always ordered from hatcheries. If you decide to go this route, you can order chicks from hatcheries from all over the country, but choosing a good hatchery will take some legwork on your part.

When choosing a hatchery, compare shipping policies, rates, and customer reviews. Ideally, you'll want to make sure you are not too far away from the hatchery you choose, so the trip for the chicks is as short as possible. You will also want to ensure the hatchery will have a consistent supply of chicks so that you can receive a new batch every two weeks during the market season.

Once you decide on a hatchery, you will want to order your chicks early enough and spaced out, so that each batch of chickens you raise will have you butchering once every two weeks. You don't want to choose a hatchery with trouble filling your order.

When you order, you will select the breed, whether or not to vaccinate the birds, and shipping dates. You'll also want to choose pullets (female), cockerels (male), or straight-run (whatever hatches) lots. Most hatcheries ship through the postal service. That method has worked well for decades, but be sure to account for possible issues with chicks dying in transit, variable arrival dates, hot weather, and other incidental problems that may crop up.

When they hatch, chicks are still digesting their yoke sack for two days, but they will need to feed after no more than three days. You'll also want to be ready to receive your chicks when they arrive. You don't want to scramble for heat lamps or other equipment at the last second.

INCUBATORS

We don't hatch our meat chicks, but we use incubators to hatch fertilized eggs (if they are not raised with the mother). Incubators allow for the egg to receive sufficient heat and moisture to promote growth, and it will usually hatch within 21 days. The chicks can then stay within the incubator for up to 48 hours before moving.

Danny's Details: It can be really enjoyable to hatch your own chickens, and we supplement our egg layers by hatching 40 or so birds every year. We pick the best and sell the rest. It's a fun project for kids, and there are plenty of hobby setups you can buy. We made our own with an old wine cooler, and then rigged a pancake-style thermostat with two light bulbs that would turn on when the thermostat dropped below the hatching temp.

BROODERS

Brooders are where you put chicks to develop after they hatch if they won't remain with the hen, or if you receive them in the mail from the hatchery. Brooders offer a heated, protected container with absorbent bedding, food and water, made to protect chicks for their first few weeks of life. For broilers, plan to keep them within the brooder for about three weeks. Brooders can be constructed from various materials, including wood, plastic, or metal.

A brooder should have two-foot walls at a minimum, keeping the chicks in, but allowing for easy access to their enclosure. You can initially get away with one-foot walls, but as the chicks grow, they may start jumping out. Depending on your situation, you will want to cover the top with mesh or grating to keep out wild birds and minimize the potential of spreading avian influenza, as well as farm cats, other potential predators, and small children. You can also use an enclosed space (such as a shed) with a closeable door. Keep your brooders in covered areas that protect your chicks from weather, including temperature extremes, drafts, and inclement conditions. Allow for six inches to a foot of space per chick. This will ensure they have enough room to move around and prevent overcrowding. Avoid any nooks or corners for the chicks to get stuck in when constructing or using a brooder. This could result in smothering, trampling, or stress deaths. Be sure to clean and sanitize the brooder before use for best results.

Here are some ideas for styles and types of brooders:

- **Cardboard Boxes:** You can certainly use cardboard boxes, but they will have the smallest lifetime of brooders and will generally be the most prone to damage. They are also the most susceptible to improperly set up heat regulators.

- **Breeder Guard:** This type of brooder is typically made of cardboard, wood, or plastic, and is set up in a square or hexagonal format on a floor. The floor can either be composed of bedding, or bare floor with bedding added.

- **Plastic Totes**: These have the advantage of being stackable, easy to move, cleanable, cheap, and readily available. They will not hold many chicks (depending on their size), but you can easily increase how many you use based on your demand.

- **Metal or Plastic Water Trough/Cistern:** Much like the totes, these are readily available, durable, and fairly portable. They come in small-ish sizes comparable to totes, but can also be much larger, which can provide more centralized brooding of chicks.

- **Bulk Breeding Cages:** These are bird cage-style breeders that are typically used to house chicks. They will come with feeders and waterers, they don't use bedding (metal grate bottoms with catch plates for waste), and they are stackable.

- **Custom-Constructed Breeder Boxes:** These can take many forms, and have the advantage of being custom-built to account for your needs and the available space. You can incorporate a closeable lid using mesh or grating, along with other features. Though these can take the most effort, they can be the best solution for your farm.

Danny's Details: When you are constructing your brooder, make sure that it is also rat-proof. One year, we built a brooder in the corner of a barn stall using a dirt floor, and a rat colony tunneled in and stole 60 three-day-old chicks. It was heartbreaking, and the potential sales of 60 birds at full-grown size in our market was $48 per bird (parted). That was a $2,400 loss within four days of start.

BEDDING

There are several types of bedding to use within your brooder, and some are better than others. The most important thing for your chicks is to give them a firm, yet soft floor that they can use to get used to walking, and something that will absorb their waste without generating an unhealthy environment. You will also want to avoid substance that can lead to slippery floors (like newspapers). On a slippery floor, chicks may be subject to spraddle leg (or splay leg) and dislocations. With meat birds that will only be alive for eight to ten weeks, dealing with leg injuries will be far more trouble than it's worth. If you're on a budget, shredded papers can work; we shredded our old tax documents and used them for a few cycles. This type of flooring soils quickly and will need to be changed often, but it's a way to save money and it composts well.

The most common and healthy bedding for chicks will probably be wood shavings. Shavings are easy to find at your local feed store, inexpensive, easy to clean out, and compost well to provide organic material for your garden. On top of that, the chicks love to scratch and move the material around their brooders. If you do go with wood shavings, there are several things to keep in mind. Don't get shavings that are too large and coarse; if the pieces are too large, they can impede small chicks making it difficult for them to walk. Make sure you get finely processed shavings that provide a smooth floor. Wood shavings have limited absorption capacity and should be cleaned out regularly.

Avoid treated shavings, as some brands use aromatic oils that are unsafe for your chicks. Cedar shavings have similar issues: the wood's natural oils are unhealthy for chickens of all ages.

Sand is another option for brooder bedding, but don't use fine particle sand, as it can cause respiratory problems. Use construction-grade sand or river sand instead. Chicks raised on sand tend to have a lower chance of bacterial disease, because the sand dries and desiccates the feces, which lowers the chances of bacteria replicating. The sand also has the added benefit of keeping chick feet healthier, and provides dust baths for them.

Sand does have its downsides though. For one, it cannot compost (although it can be mixed with clay soil to provide some drainage to your garden or fields). Sand also easily compacts, and can develop into a hard surface unsuitable for your birds. It can become very warm during hot days or under a heat lamp.

Straw is the third option, but it is the least favorable of the three. Straw is inexpensive, natural (as long as it's untreated), and can provide extra insulation during colder conditions. In the limited area of a brooder, straw is not overly absorbent and can tend to mold rather than compost if not cleaned regularly. This can lead to respiratory diseases for chicks. Larger pieces of straw can also make it difficult for young chicks learning to walk. It can, however, be a good option after chicks have moved out of brooders but aren't on the field yet.

HEAT

When you're preparing your brooder, you'll want to provide heat to the enclosure for at least 24 hours before your chicks arrive. Newly born chicks cannot regulate their body temperature, so they are at high risk from improper or insufficient heating. Not only do they require heat, but without their mothers they will need to be able move freely in and out of heated areas to find their ideal comfort zone.

Be aware that you will not want to heat your brooders uniformly. You need to provide different zones for the chicks so they can move around and regulate their temperature. Heat lamps should be suspended above the brooder so that you can raise or lower them to create a circle of heat, then adjust it depending on the behavior of your chicks. If they are too cold, they will cluster together and huddle under the heat, which can cause smothering and death; if they are too hot, they will spread out away from the light. You should also listen carefully to the chicks. If they are making sharp, loud cheeping sounds, it means they are not happy or comfortable. Keep an eye out for chicks that are lethargic, or don't have an appetite. Check for legs that are puffy, swollen or cold to the touch. Happy chicks will make soft peeping sounds that don't sound distressed.

Dannys Details: Be sure to pay close attention to how you rig your lamps. We once had a light fall and break, and it nearly caused a fire. We were very lucky. Our seven-year-old happened to see it smoking on the ground, and buzzed us on the intercom. We may have set personal speed records as we sprinted down the hill to the barn in a near-panic. The moral of the story is that you should make sure any lamps are properly secured. If your lamps aren't secure, all it takes is an animal bumping into it, and you will suddenly have a very bad day.

Here are the heat levels to keep your chicks at until you can move them outside to their grow-out pens or tractors.

Age	Cage Brooding	Floor Brooding
Day 1-3	33-34°C or 91-93°F	35°C or 95°F
Day 4-7	32-34°C or 90-93°F	33°C or 92°F
Day 8-14	29-31°C or 85-89°F	31°C or 89°F
Day 15-21	26-29°C or 80-84°F	29°C or 84°F
Day 22-28	24-26°C or 75-79°F	26°C or 79°F
Day 29-35	21-23°C or 70-74°F	23°C or 74°F
Day 36+	21°C or 70°F	21°C or 70°F

HEAT SOURCE OPTION

Here are a few options for the best ways to generate heat and avoid drafts:

Brooding plates (hover heaters): These mimic the mother hen by providing a close, covered area that can also give chicks places to run, hide, and keep themselves warm. Brooding plates should have an adjustable height to accommodate growing chicks. They should also have separate front and back sections that can each be adjusted to offer a heat gradient. Most heat plates resemble a small step stool or plate suspended between four legs. Many farmers use these in conjunction with heat lamps. Suitable for small farm operations.

Heat lamp: Probably the most common form of heat source for chicks on small, independent farms. Heat lamps are inexpensive, easy to find, simple to use, and can be adjusted depending on your setup. They are very hot on their surface, and should avoid touching anything flammable. White bulbs tend to be more of a fire hazard; you should also use infrared heat lamps with red bulbs so your chicks can settle into a natural circadian rhythm. Avoid food heat lamps, as they are coated in polytetrafluoroethylene (aka teflon), which can release toxic fumes. Most heat lamps use a 250-watt bulb. That can generally accommodate up to 100 chicks for a single bulb. If you're aiming for somewhere between 300-500, a four-bulb setup should suffice. Suitable for small farm operations.

Forced hot-air heaters: Powered by fuel of various types, or electricity, these machines heat the air and distribute it to the poultry shed. They use interior air, requiring increased ventilation to avoid any build-up of airborne toxins. Forced air heaters can serve a large area, but inefficiently heat the floor. Suitable for large farm operations.

Radiant heat: These devices combust gas to heat radiant surfaces, and can take the form of angled reflectors, tubes, or disks suspended above the floor. Radiant heat is a popular option, because it delivers heat straight to the birds and to the floor, plus it is more efficient than forced air heat. They are also not considered fire hazards. Suitable for large farm operations.

WATER

Ensuring that you have a consistent access to water sources is crucial for your chicks' health and well-being. Keeping that water clean is equally important. Several options exist depending on the size of your brooder and the number of chicks you want to raise. One common recommendation is to mix in about one tablespoon of apple cider vinegar for each gallon of water for your chicks. Along with helping with overall health, there are potential benefits for the animal's digestion.

Try to avoid providing your chicks with water in an open dish or saucer. This can contaminate the water with bedding and feces, and could result in your chicks getting wet, or even drowning. If the water is deep enough to cover the nose holes in the beak, there is a chance that the chick will inhale water into its lungs, which can lead to drowning.

There are a few different types of waterers, though the most common form is with a container fitting upside down into a tray (not unlike an office water cooler). This configuration allows water to flow down and out as needed, without overflowing the tray. To avoid drowning risks, put marbles or river rocks into the tray to allow the chicks access to the water, provide weight stability, and limit the depth in which the chicks can sink their beaks.

When you're starting out in a brooder, a one-quart waterer will serve about 25 chicks. Move up to a one-gallon container when the chicks have grown a bit, usually when they are big enough to repeatedly empty it, and when they are boisterous enough to knock the one-quart container over. You may find it easier to put the waterers onto a board or tile for stability in the brooder. That makes them easy to use, but also makes them prone to being knocked over and spilled.

Another very successful option for waterers is a nipple system. This usually takes the form of an elevated container or pipe system, with suspended nipples hanging down within the chick's reach (keep the nipple height between where the chick's eyes and back are for best results). The nipple is usually a piece of metal or plastic, which the chick pecks, releasing a small amount of water that they drink. This system can also be scaled from small to larger coops with no risk of drowning.

Another option combines a nipple system with a small cup, which provides water in a dish, but not in an amount that will drown a chick.

Waterers should be cleaned and checked regularly to ensure they are in good working order. Keep the water containers full, so water is always available for your chicks.

Danny's Details: One year, we had a leaky waterer that poured out into the brooder. The bedding got wet, and in turn, birds got wet and then cold. That led to huddling, and many of them got smothered under the huddle pile. Make sure the waterers are not leaking, and that they are tightly screwed onto the drinking trough before you leave the brooders alone for any period of time.

FEEDERS

Broiler chickens are bred for rapid growth, and encouraging that growth requires a high-protein diet. Underfeeding them, or providing inadequate access to feed, can result in stunted growth and poor overall health. Additionally, feeding competition can cause stress and aggression among birds, leading to injuries and even death. To prevent underfeeding and competition, ensure that all birds have equal access to sufficient quantities of feed, and that the feed is of high quality.

Having the right feeder to distribute the food evenly to your chicks is also important. Chicks are messy and will kick, step on, and climb on whatever kind of feeder they have. This can result in plenty of waste. Food can be kicked out into the bedding and potentially lost, and if the chicks stand in their food, they will eject their waste into it, causing contamination. That means it's important not to use just a plain dish or open trough.

Most feeders designed for more than a few birds work to eliminate these issues by providing multiple size-restricted holes for the birds to feed through, and it won't allow them to enter the feeding container. These feeders also have methods to limit perching on top of the container, so the birds cannot eject their waste into the food.

Most feeders come in either metal or plastic. Plastic is cheaper, but it won't last as long as a good quality metal product. Sun, cold, and other environmental conditions can also harm plastic, but it's easier to clean than metal.

One of the most common forms of feeders is a long, plastic or metal trough feeder. They are inexpensive, easy to find, and easy to work with. While chicks are small, they can feed side by side, and they are designed so that the birds cannot climb into the container. That encourages immediate feeding, and the covered top keeps the chicks from kicking food around. Many of these troughs also have an angular top, which discourages chicks from perching on the top (and pooping into the food, but it doesn't entirely eliminate it. Another option is a roof peak featuring rotating bars, which does a better job of keeping birds from perching.

The main issue with trough feeders is that they are fairly lightweight, and as the birds grow, they can kick them over and spill the food. Small troughs are best used when the chicks are small, then when they have spent a few weeks growing, you will probably want to upgrade to a larger, heavier feeder. Trough feeders also require being topped off frequently due to their limited size, although there are multiple options of varying lengths to accommodate different-sized flocks, from small to large.

Danny's Details: We used old gutters for troughs, and fastened a piece of plywood on each end on our grow-out tractors to keep them from tipping over. Having long feeders as they age is a good option to prevent overcrowding and bullying, something we saw frequently with round dispensers.

Mason jar, or hopper feeders are another option. They contain more feed so they don't need to be topped off every few hours, but they also have less feeding area compared to long feeders. That makes them less efficient for large flocks. They consist of a large container turned upside down into a dish, allowing the food to spill at a measured pace as the chicks eat. Mason jar feeders are limited by the size of their base, so they don't scale as well as trough feeders.

Hanging, or bell-shaped feeders are good at preventing chicks from climbing into, or perching on the container. They require something to hang from, so if your brooder doesn't have a top this may prove difficult. You can also secure a board, pipe, branch, or something else set atop your brooder to hang the feeders; just make sure to attach the feeder securely or it might be pulled down. You can also adjust the height of the feeders depending on the age of the chicks. A good rule of thumb is to hang the feeders so the feed is at the height of the smallest chicken's back. Hanging feeders also help deter mice or rats among older chickens.

There is also an option to use automatic, tread-plate, and treadle feeders. In general, they aren't good for chicks or meat birds, partly because there a steep and lengthy learning curve. They are also much more expensive than other types of feeders.

LIGHT CYCLE

A light source in the brooders is important for chicks raised without mothers. Leaving the light on does not disturb their sleep, and will provide a sense of security that helps combat their poor night vision. As mentioned in the heat section, if you use heat lamps, you should use red bulbs vs. white ones. Using 24 hours of white light can cause stress for the chicks, resulting in pecking behaviors and other negative issues. Chickens are attracted to the color red and tend to peck at it, so if a chick is hurt, or has a bit of red umbilical hanging from its vent, it could attract its siblings to peck at it. Red light camouflages this and will keep the chicks calmer.

If you use a heat source other than a lamp, and the brooder is in a place where natural light is minimal, you should provide cyclical lighting. Regulated lighting levels during the daytime, and a night light are good in helping to establish a natural day/night cycle for the chicks. Do not turn off the lights abruptly; the sudden darkness will cause chicks to panic and could result in stress, huddling, and trampling.

DISEASE, SICKNESS, AND PARASITES

Upon receipt of your chicks, it is important to do an initial check for sickness or diseases, and you'll want to periodically check your birds throughout their lives. It is important to isolate and remove a sick chick from your flock and treat or cull them as soon as possible. Recognizing illness in a bird or flock as early as possible will give you the best chance to prevent loss.

As a prey species, chickens will instinctually hide their illness to avoid standing out to predators. They can even pretend to eat. Signs to look for are birds that stay by themselves, are lethargic, and have discoloration or discharge from the nose, eyes, or vent. Inspect the eyes for cloudiness or bubbles. Listen to breathing for any whistling or distress. Inspect the feet and legs for swollen areas or sores.

Because these are meat birds, it can be necessary to determine the cost of treatments vs. the bird's life. Culling the birds to protect the rest of the flock may be the better option.

COMMON ILLNESSES

Aspergillosis (brooder pneumonia) is caused by fungus spores that develop in warm, moist, dirty environments, such as an ill-maintained incubator or brooder. Aspergillosis is not spread between birds, only environmentally. Chicks are especially vulnerable because the new cilia in their throat aren't matured enough to move the fungus spores up and out. Symptoms include open-mouthed breathing and gasping, nasal discharge, slow eating, and increased drinking. They may also have nervous system symptoms such as tremors, inability to balance, and head twisting. Symptoms may look similar to Marek's disease, and are typically diagnosed by microscopic evaluation of the fungus taken from the internal respiratory system. The best prevention is to keep everything clean, and remove wet litter. There are treatments when chicks become ill, such as nystatin and amphotericin B, but they are expensive. Brooder pneumonia can cause a death rate of up to 50% of a flock. The spores can infect humans, as well.

Coccidiosis is caused by an intestinal parasite from cocci eggs found in chicken feces. Because the birds peck at anything, they can ingest these eggs, which then hatch and burrow into the chick's intestinal wall. This causes some bleeding, characterized by an orange or red color in their poop, which may also be frothy and contain mucous. Symptoms can include dehydration, withdrawal, and limited eating. While your chicken may survive without treatment, they will probably never be as healthy or productive. The best way to prevent coccidiosis is by changing out bedding often, and keeping your coop or brooder dry. You may also try a medicated chick feed (they do not contain antibiotics) that reduces the number of harmful protozoa in their gastrointestinal tracts, until they develop their adult immunity. There are different strains of coccidian, and your birds may be infected multiple times, especially in stressful or changing environments.

Infectious bronchitis comes from a type of coronavirus with several subtypes. Symptoms may resemble a human cold with heavy breathing, gasping, mucus discharge from the nose or eyes, coughing, depression, sneezing, and huddling together. Infections can spread through air, food, and contact. The cold will spread rapidly, so if you cannot isolate the infected chick quickly, all the birds will most likely end up infected within a few days. This hits chicks under six weeks of age the hardest, and can lead to high mortality rates. There are vaccines that can help prevent infectious bronchitis, but they won't help if an outbreak is already underway. The prevalence of subtypes and mutations also makes it difficult to prevent it completely. The best treatment is to raise the temperature 3-4 degrees, and provide antibiotics. Chicks sick with a cold are susceptible to secondary infections as well, so keep them clean with good food and water.

Marek's disease is a viral disease that is almost always fatal. Most hatchery chicks are vaccinated against it in their first 24 hours after hatching, or sometimes even while they are still in the egg. If you hatch your birds, you should consider vaccinating your day-old chicks as they respond less and less to the vaccine as they age. Most chickens have probably been exposed to Marek's without becoming ill at some point, but being stressed can weaken their immune system enough to catch it. Marek's has a two-week latency period while still contagious before the chick becomes visibly ill. In chicks, it typically manifests as weight loss, leading to death within about eight weeks. Older chickens have other symptoms, such as cloudy eyes, leg paralysis, and tumors.

Omphalitis (mushy chick disease) is usually caused by an infection to the navel soon after hatching, or caused by improper egg washing pushing bacteria into the shell. Chicks may even die before hatching. Symptoms may include unabsorbed yolk sacks, dehydration, and a putrid smell. The abdomen may be distended, and chicks can be lethargic, huddling near a heat source. Omphalitis may be caused by poor sanitation in the incubator or brooder, a chick pecking at another's navel, or a handler confusing the navel scab or dried umbilical cord for pasty butt, and attempting to clean it off. Prevention is cleanliness, not incubating dirty eggs, applying some iodine to unhealed navels on your chicks, and removing any infected ones.

Pullorum (salmonella) comes in many strains. Some are well known because they are dangerous to humans, but those strains are usually different from those that can be harmful to chicks. Symptoms may include light-colored diarrhea, fatigue, loss of appetite, shriveled/purple comb (when older), pasting, and wattles, all leading to death. A conclusive diagnosis is typically post-mortem from lab identification of bacteria. Some antibiotics have been shown to eliminate salmonella enteritidis – the strain carried by chickens that can be dangerous to humans – in very young chicks (one week or less in age). While antibiotics may effectively treat a sick chicken, salmonella can be latent and remain infectious. Some salmonella strains must be reported to health authorities. The bacteria can survive on cast-off feather dander for five years, transmitted directly into an egg by the hen, infected droppings of other chickens or rodents, or contaminated equipment. It is best to eradicate any infected birds immediately (sometimes required by law, depending on your area), and be sure to buy from clean, trusted sources.

Necrotic enteritis (rot gut) is an illness that produces rotten-smelling diarrhea and listlessness in the affected chicks. It is a bacterial infection that typically spreads through overcrowding. Antibiotics administered in water can treat infected chicks, but the best prevention is proper cleaning and not overcrowding.

Avian influenza (bird flu) is a contagious respiratory illness caused by influenza viruses. There are two types: LPAI and HPAI. LPAI is the less contagious form and is very common. HPAI, the "highly pathogenic" version, is a serious issue that can potentially destroy your entire season, and potentially your livelihood. It may also lead to federal authorities becoming involved. That could lead to culling every bird on your property and destroying structures.

On rare occasions, some strains of bird flu can jump to humans, but so far H5N1 is the only type that has been able to infect people. Symptoms for the birds impacted by H5N1 may range from decreased egg production (not a concern for meat birds but worth noting), breathing difficulty (coughing and sneezing), diarrhea, hemorrhages in the feet or legs, lethargy, reduced food and water intake, twisted neck, paralyzed wing, or even sudden death. It can be spread through saliva, nasal secretions, poop, and contact with virus-contaminated surfaces.

There is no cure or current vaccination for the bird flu, but there are ways to limit the risks of contracting it. To prevent bird flu in chickens, house your birds away from where waterfowl congregate (wild migratory waterfowl are the primary spreaders of the virus), and limit exposure to wild birds. Create shelters with sturdy roofing materials and cover them with poultry netting to prevent wild birds from roosting on top of your brooder or shelter, which can lead to contamination via their droppings. Keep food and water inside the brooder or shelter to prevent wild birds from accessing these resources. Regularly clean and disinfect your shelters and brooders.

Danny's Details: In 2014, we were preparing to move south to the new property when a bird flu outbreak started raging across the country. It was estimated that 59 million birds were infected by the time it was over. We only had about fifty birds, so we didn't initially think much about it. Little did we know we were about to be in the thick of it.

In December, we were getting scared. The H5N2 virus was in the news, and there were major losses being reported across the Pacific Northwest. The USDA was taking it very seriously, and we started receiving FAQs and recommendations telling us to keep farm birds away from migrating waterfowl paths, and areas where they may bed down and leave droppings. Being on the Olympic Peninsula, we were in a major flyway and resting spot for birds migrating south from Canada. The weather had turned wet and cold, a typical winter in the Pacific Northwest, and the first reported case of bird flu in the U.S. was in Oregon. It seemed far enough away that we didn't worry too much about it, but with the move coming, and more migratory birds flying over our property and landing in our fields, we started keeping our birds cooped up in the barn to lower any of the risks as much as we could. One of my least favorite memories from that experience was capturing the domesticated call ducks that used the outdoor stock pond and putting them into a shelter (our dog really enjoyed it, but we didn't).

Sharon moved to the new property first, and within three days the first outbreak hit Washington state, just a few hours away in Benton County. With rumors of a possible county quarantine, we advanced our timetable, and started moving the flock after the first of the year. We were friendly with many other poultry farmers on the peninsula and planned to grab a few more turkeys, but we hadn't had time to grab them yet. Our friend was going to sell us a few turkeys, but she decided stop all sales until she could test her birds. A few weeks later she lost her entire flock to H5N2. Her turkeys had to be culled, and all the structures associated with them were destroyed.

By January 21, all of Clallam County (our home county) was quarantined, and forbidden from moving birds. Thankfully, we made it to Oregon about two weeks earlier, and three days before our friend's flock tested positive. It was pure luck that we didn't get it. I still feel so bad for her losses, but I'm also grateful that she showed responsible animal husbandry by not selling us birds when she was unsure. It was a devastating loss for her business, but it may have saved our future.

When we learned of our friend's calamity, we went into full biosecurity mode. We kept our birds in covered areas in the barn for the better part of two months. That resulted in some seriously un-fun times keeping everything clean. We also incorporated new safety procedures, like bleaching our boots going in and out of the barns, and we didn't allow anyone onto the property. Hand washing and leaving boots outside were tantamount. It was an eye-opening experience. Overall, $1.1 billion dollars of broiler exports and $41 million in egg exports were lost between 2014 and 2015, along with more than 51 million birds.

We managed to avoid H5N2 at the new property, and those lessons helped us deal with COVID as well. Doing your part in stopping the spread of viruses is an important responsibility, and something you should consider before starting a poultry operation.

Newcastle Disease (vND) is considered one of the world's most significant poultry diseases. There is no treatment, and it is highly infectious. It can be transmitted through the inhalation or ingestion of the excretions from infected birds. It can also be associated with contaminated food, water, and equipment. The best treatment is keeping your bird's environment clean, dry, and well-maintained. There are vaccines available as well.

Avian encephalomyelitis has no cure, and infected chicks must be disposed of. Symptoms are dull eyes, head and neck tremors, limited feeding, and eventual paralysis. A vaccine is available to prevent the disease, but for meat birds it might be prohibitively expensive.

Fowlpox can result in wart-like growths or sores, slow growth, and mouth cankers. Infected mosquitoes often spread the virus, and prevention can be accomplished by spraying for mosquitoes. There is a vaccine as well.

Pasty butt, also called pasting or pasted vent, occurs when feces stick to the anus and blocks it off. This can cause gut impaction and constipation. The cure is to keep the birds clean and unblocked.

Airsacculitis, also known as air sac disease usually only occurs in chicks, and can be caused by E. coli bacteria, or mycoplasma gallisepticum. Symptoms include weight loss, coughing, nasal discharge, watery eyes, and difficulty breathing. To avoid this disease, stay away from dusty litters, maintain good ventilation, and keep chicks warm. If chicks become infected, feed them a high-protein diet with plenty of vitamin E. Be aware that survivors will be carriers that can affect other birds.

Breast blisters are most commonly associated with Cornish Cross breeds, but can become an issue with other birds that are reluctant to move around much, or have arthritis. This can also affect birds raised on wire or hard dirt. Keeping your chicks and chickens on soft bedding is the best way to avoid this condition. Breast blisters are pretty much as they sound, with fluid-filled blisters developing along the breast bone of a chicken. If the fluid remains clear, it may resolve itself, but if the fluid becomes yellow or cheese-like, determine the underlying cause and treat for that.

Blackhead is caused by protozoa, which is spread by worms in chicken droppings. It's mainly an issue with turkeys, but chickens with depressed immunity can be infected, and all chickens can be carriers. Symptoms include lethargy, droopy wings, loss of appetite, weight loss, and stunted growth. Bloody or yellow-ish droppings can also be an indicator. You can use medicated chicken feed and dry their living conditions to treat.

Bumblefoot is a staph infection in the foot. Symptoms include limping and a black scab on the bottom of their foot. To treat, you should clean the abscesses, use antibiotics, and move the chicken to a clean place.

Chronic Respiratory Disease (CRD) is caused by mycoplasma bacteria. It comes on slowly, and generally affects older birds. Symptoms can include weeping eyes, which can lead to the chicks' eyes being stuck shut. You may also notice difficulty breathing and nasal discharge. To prevent CRD, make sure your birds have plenty of ventilation, and keep them warm. Isolate any infected chickens from others.

Infectious Bursal Disease (also called Gumboro Disease) is caused by a virus in the lymph tissues. Sometimes chicks can be born with maternal antibodies leading to no symptoms, making it difficult to tell if they have been affected. Early signs are white diarrhea that stains the vent feathers. You may also see lethargy, reduced appetite, difficulty walking, or reluctance to stand. There's no real middle ground with this, and infected birds will either die or recover. There are vaccines, but they are not widely used.

Laryngotracheitis is a relative of the herpes family, and causes an upper respiratory infection of the larynx. Symptoms include watery eyes that turn yellow and crusty, coughing, sneezing, or stretching their necks out while breathing and shaking their heads. No treatment exists, and any surviving birds can infect others, so starting with a new flock is best. Thoroughly clean everything and wait two months to avoid reinfection.

Intestinal worms can come in various forms, like roundworms, tapeworms, capillaria, and cecal worms. Symptoms can include diarrhea, poor absorption of nutrients, enteritis, rough feathering, stunted growth, pasty vents, and pale birds. Follow directions on the various treatments on the market.

Parasites can come in the form of either lice or mites. Check between the fluff and feathers at the base of the tail and vent area. If you see tan, oblong bugs, your birds have lice; if you see black specks, your birds have mites. There are plenty of treatments on the market.

Now that you're properly terrified by all the invisible things that can kill your birds through a variety of horrible symptoms, including bleeding from their anuses and a vicious strain of herpes, let's talk about adorable box chicks.

Once you have a box of chicks and they are settled in, take a half hour to watch them. Ensure that everyone is eating and drinking, nobody is getting picked on, and there are no weird issues. When you tend them, take a little extra time each day to watch them to guard against loss from competition, leaky waterers, huddling, etc. If you want to just throw feed and water in then go about your day, you are in the wrong business. Tending animals is challenging, but it's also fulfilling. Remember that life is precious, and how your birds are treated ultimately affects you. Tending to your birds is also tending to you.

4

Grow-Out Pens, Pasture Tractors, and Finishing Tractors

Keeping your chicks healthy and safe and allowing them to grow is your goal, and at each stage in their life you'll need to utilize a variety of tools. Perhaps the most important "tool" for raising pastured poultry is the "Chicken Tractor." Many farmers call their mobile chicken tractors different things, and there are thousands of ways to make them.

The way we use these mobile structures is a hybridized variation on some standard styles to achieve the 1,000 bird in one summer method. We break them into subgroups that we call "grow out pens," "pasture tractors," and "finishing tractors." These are tractors I created with different heights and sizes in mind for the growth phases of our birds as part of our 1,000-bird method.

GROW-OUT PENS

Everything we do within the 1,000-bird method is broken into fours. You will need to create four brooders that can contain 25 birds each. You'll want to have four grow out pens (or two large ones), four field tractors, and four finishing tractors. Four is the magic number. I even started breaking down the different stages of a bird's life into four different levels based on their maturity. That lets me butcher a new set of birds every two weeks.

Using grow-out pens for meat birds is usually a simpler process than with other types of birds. You don't need to socialize these chickens with an existing flock, but you also don't want to expose them to the full elements. Grow-out pens commonly have solid, or raised wire-mesh floors, designed to be cleaned and sanitized regularly. We generally use eight-foot diameter stock tanks with two-foot walls for our process, which is enough room to provide for about fifty juvenile birds. There are plenty of options to buy other types and sizes of pens, but if you construct your own, you can use wood, PVC pipe, and metal materials. Whatever option you prefer, choose sturdy, durable, and easy to clean material.

Danny's Details: Grow-out pens are typically used to introduce chickens to a flock, by allowing the birds to safely socialize within, or next to an existing yard. For meat birds, however, you won't want to introduce them into an existing flock. We use grow-out pens to wean them off artificial heat, and get them acclimated to an outdoor environment as they feather out. Play this by ear, as you may need to provide heat longer early in the season.

Here are some further guidelines to follow:

- Grow-out pens should allow for between 1-2 square-feet per bird. This allows the birds to move around comfortably and helps to prevent overcrowding, which can lead to health problems and increased stress.

- Secure your grow-out pens against threats and predators by constructing solid or wire/plastic mesh walls and roofs. Mesh will allow you to monitor your birds more easily, but make sure the materials is securely fastened to the pen frame, and the openings are small enough that predators can't enter. Any door or gate should also be secure against predators and other intrusions.

- Make sure that you have good ventilation to protect your birds from direct drafts. Add fans or vents when necessary to prevent respiratory issues and the build-up of gasses like ammonia.

- Use the same bedding strategy as a brooder. Provide dry bedding in the pen that absorbs moisture and prevents bacterial growth. Suitable bedding materials include straw, wood shavings, or shredded paper. If you smell ammonia, change out the bedding immediately.

- Use larger, heavier feeders and waterers. The bigger your chickens are, the more food they will require, but that also means that they can knock over your equipment if it is too small or light. Food and water should be provided at a level that is easy for chickens to access, usually at a height located between the beak and shoulder blades. Ensure there is enough area for both the feeders and waterers to avoid aggressive competition.

- Choose a well-drained location away from areas where water may pool or collect. Avoid areas with heavy foot traffic, and consider the proximity to water and electrical sources.

- If the grow-pen isn't already indoors, you'll need to provide shade and shelter for the birds to protect them from the elements. Ensure that part, or all of your pen is covered in some fashion.

- Your birds should be fully feathered at this point, but keep an eye on their heat requirements. If it is cold, provide supplemental heat to help them transition successfully.

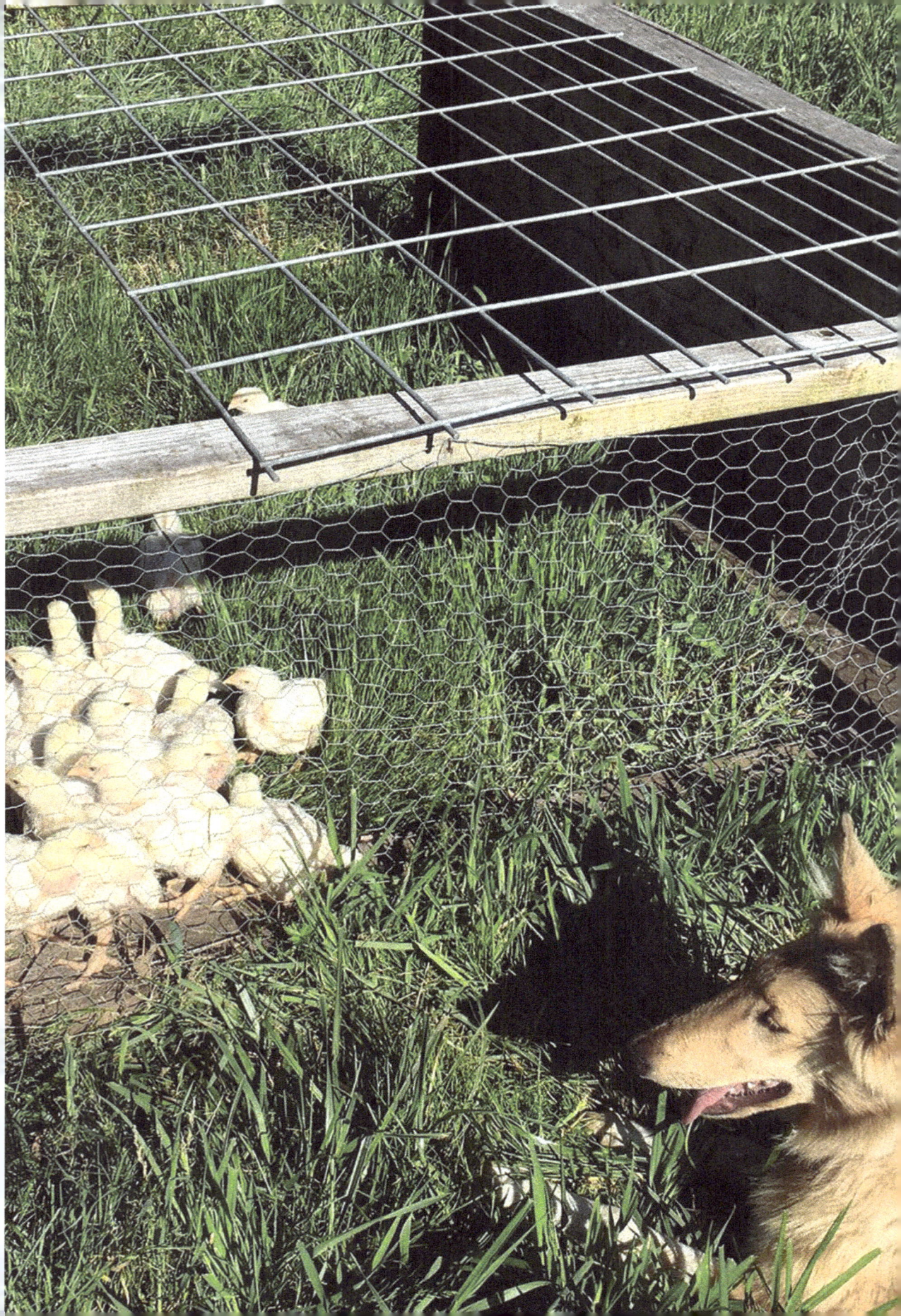

PASTURE TRACTORS

In contrast to grow-out pens, a pasture tractor is a movable structure (coop) designed to move around a pasture or field. It typically has an open bottom, allowing the chickens to graze on the grass and forage for insects and other natural foods. Pasture tractors are often used for raising chickens for meat and egg production, and can provide a more natural environment for the birds than a stationary grow-out pen. Grow-out pens and pasture tractors both have advantages and disadvantages.

Grow-out pens provide a regulated environment for raising chickens, with greater control over temperature, ventilation, and lighting. On the other hand, pasture tractors provide a more natural environment, with access to fresh grass and other natural foods. We use both methods, using grow-out pens for juvenile chicks, and then moving them to pasture tractors once they are older.

🔨 BUILD YOUR OWN TRACTOR

If you're building a chicken tractor, you'll want to factor in both what you'll need for movement, and who will be moving it. If you're moving the tractor by hand, several smaller tractors will probably be better than one huge tractor that can't be moved! Chicken tractors can be moved manually, or they can be attached to a vehicle like an ATV or a motorized tractor for easier transport. Attaching wheels to one end, using a lift dolly, or attaching ropes on both ends for lifting are also options.

The choice of roofing material on a chicken tractor is also important. Aluminum roofing is lightweight and durable, making it a popular choice, but steel or polycarbonate roofing can provide better insulation in colder climates.

Tools
- Circular saw
- Drill
- Hammer
- Staple gun
- Power screwdriver
- Tin snips or wire cutters
- Tape measure
- Level
- Square

2×4 Pressure-treated lumber
- Eight-foot board (x4)
- Six-foot board (x6)
- 24-inch board (x10)

Plywood
- 2×6 ft (x1)
- 4×6 ft (x1)
- 2×4 ft (x2)

Additional Items
- Corrugated metal or polycarbonate roofing cut to four-by-six-foot
- Heavy-duty wire poultry netting or mesh, 2×12 ft Cattle panel cut to 6×4 ft
- 10-foot tow rope
- Fasteners (staples, nails, screws)
- Paint or stain (optional)

1.

Begin by designing and "laying-out" your materials on a clean, flat surface, like a concrete pad or garage floor. The coop's size will depend on the number of chickens you intend to keep, with 1.5 to 2-square feet of space per bird being the recommended minimum. Our 48-square-foot tractor design, for example, can hold up to 25 birds comfortably.

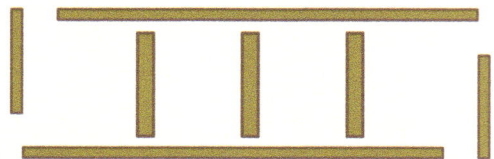

2.

Screw the top and bottom pieces of lumber together on each end with a stud, ensuring a flush fastening plane for the wire and sides. Space a stud at two, four, and six-foot intervals, resulting in a pair of two-foot-tall stud walls that are eight-foot long, with a four-inch-wide bottom skid, and a four-inch wide top

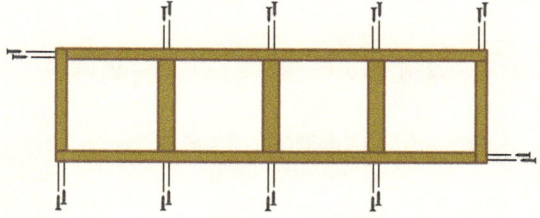

3.

Take two six-foot 2×4 (spanners) and lay them on top of the bottom 2×4 wall skids, creating a bridge between the two walls on each end, with a small gap the width of the lumber under these ends. Fasten the top spanners in line with studs, as shown in the illustration, to finish the box frame.

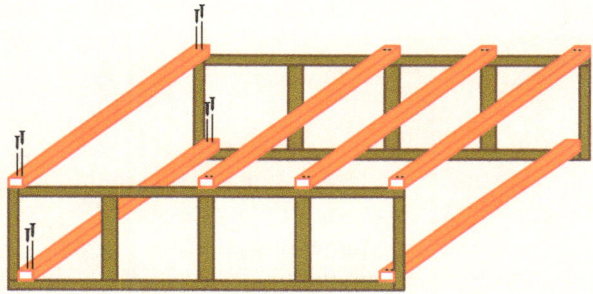

4.

Add plywood or exterior-grade sheeting to the top, back and front of the box frame. On the six-foot side, this is where your six-foot plywood will fasten. Attach both of your four-foot pieces to the corresponding stud walls in the illustration. Then put the 6×4 ft roof plywood on. The plywood pieces form the "safety box," where birds can roost or huddle at night without worrying about predators. Attach these pieces to the frame using screws, and then fasten small plywood triangle gussets to the front corners where there is no plywood to strengthen the tractor.

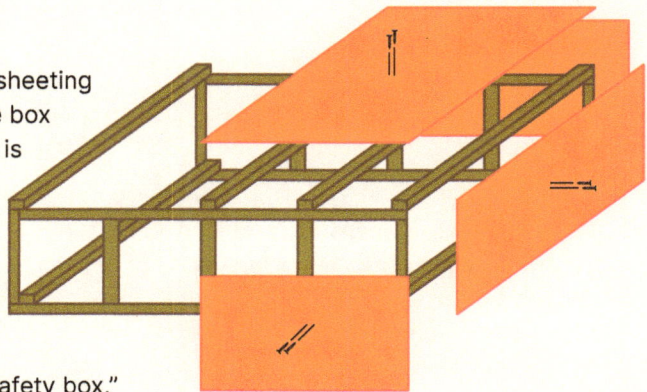

5.

Cut and fasten the hardware cloth or poultry netting to the sides where needed. At this point in the build, your tractor should be fenced in on all four sides. Make sure you use enough fasteners to keep the netting tight so it's not pry-able by little raccoon hands, coyote noses, or owl beaks!

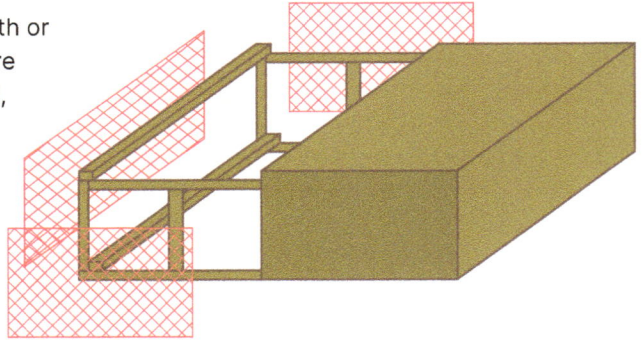

6.

Cut your polycarbonate or corrugated roofing to fit the 4×6 ft roof area and fasten to the plywood. This will give protection to the plywood underneath, and still get structure from the box. If you want to make the tractor slightly lighter, you can remove the plywood top piece and just fasten the roof material by itself, but this will give the tractor less structure and a shorter life span.

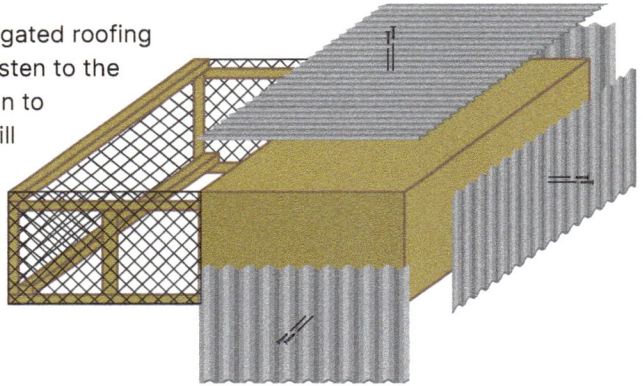

7.

Take your cattle panel and make sure it is cut to accommodate the remaining open area of the chicken tractor. This should be about a 4×6 ft open area. Make sure the cattle panel is cut to cover this opening.

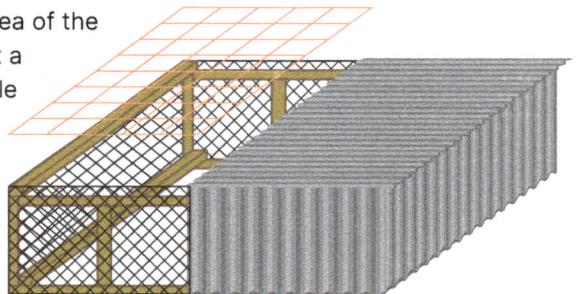

8.

Add poultry netting to the cattle panel frame with wire of zip ties. This may sound excessive, but smaller birds of prey have no issue dropping through the larger openings of a cattle panel for a quick meal. Once your cattle panel has the netting fastened to it, put it in place over the opening. Go around it and confirm that there are no holes that need additional attention, and that the netting is secure. Your tractor is complete!

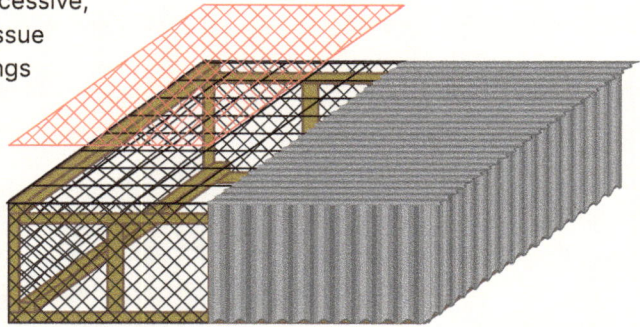

There are hundreds of designs for chicken tractors, but this basic wood-framed, open-bottomed design is a great starting point.

ROTATING PASTURES

Rotating pastures using a pasture tractor (not a regular coop) is necessary while raising chickens for meat or eggs. Rotating pastures involves moving the chicken tractor to a new location every few days or weeks, depending on the size of the pasture and the number of chickens. This allows the birds to access fresh grass and bugs while preventing the buildup of waste and parasites in a single area.

The specific ratio of chickens-to-tractor size, and the frequency of movement can vary based on several factors, including the chickens' age, breed, density of vegetation in the pasture, tractor size, number of chickens, and size of the pasture.

Keeping in mind that you should allow 1.5 to 2-square feet of space per chicken in the tractor, you should move the tractor to a new location every 1-2 days for 12 chickens or more. To fine-tune the movement times, you'll need to observe your chickens and their behavior to determine if they need to be moved more or less frequently. If the chickens are scratching and grazing vigorously, it may be time to move the tractor to a new location. Conversely, if they are spending a lot of time resting in one area, you may be able to leave the tractor in place for a few extra days.

The size of the pasture is also a factor when determining the movement frequency. If the pasture is small or has a high vegetation density, you may need to move the tractor more frequently to ensure that the chickens can access fresh grass and bugs. Additionally, if you overgraze the pasture, it could lead to soil erosion and reduce the land's overall health. To prevent overgrazing, you may want to divide the pasture into sections and rotate the chicken tractor between them, giving each area time to rest and recover. This can also help to ensure that the chickens have access to fresh grass and bugs in each section of the pasture.

The benefits of rotating pastures using a chicken tractor include:

- **Improved Soil Health:** The chickens' grazing and scratching behavior can help aerate the soil and add nutrients, leading to healthier pastures and better plant growth.

- **Reduced Parasite and Disease Pressure:** Moving the chickens to a new location regularly can help break the parasite and disease cycles that can build up in a single area.

- **Higher Quality Eggs and Meat:** Chickens with access to fresh grass and bugs can produce healthier, more flavorful eggs and meat.

- **Reduced Feed Costs:** Chickens with access to fresh pasture can supplement their diet with grass and bugs, reducing the need for expensive commercial feed.

Overall, rotating pastures using a chicken tractor can be an effective way to raise chickens for meat or eggs, while also improving the health of the land.

FINISHING TRACTOR

Finishing tractors are similar to chicken tractors, but are used for the final growth stage before processing (6-8 weeks for Cornish Cross). If you add roost bars to your enclosure, make sure they are only four inches or less off of the ground, as broiler chickens are very heavy at this point and can easily break or dislocate their legs and wings. Roosts can become soiled with feces, leading to health issues for the birds. Ensure that the roosts are easily removable and replaceable for regular cleaning. Mine were just made by screwing a 2x4 on flat to a wider 2x4 leg on flat. They are easy to remove and sturdy. Our finishing tractor at five-feet-tall lets you stand up and remove the chickens far easier than if you have to crawl into a two-foot-tall pasture tractor enclosure. When you are processing 50-100 chickens in a day to take to the farmers market, this will make a huge difference.

To reach the goal of having 1,000 birds in a summer, and to do it with only two people, you'll need to use graduating tractors to establish a constant cycle of birds.

Big Al, our electric converted tractor pulling a finishing tractor onto new grass.

PREDATOR DETERRENCE

While raising chickens in the pasture is far healthier and better for the birds, it has its risks. Predators are always present, both in the sky and on the ground. Keeping chickens safe from other animals during every stage of raising your chickens is something you'll want to consider as you build your various tractors. Predators will go to great lengths for a meal at your expense, and the results can be disheartening. Foxes, coyotes, hawks, rats, and many other types of predators are a threat at all stages, but there are ways you can combat these dangers.

Start by placing the chicken tractor in a secure location that is not easily accessible to predators. Avoid heavy vegetation, and an overabundance of natural cover. You don't want predators to feel safe sneaking up to your tractors! Use a sturdy design, with material that cannot be easily broken, pried open, or damaged. Use heavy metal mesh or grating to resist damage, and ensure the frame is not weak. You will also want to make sure that your doors can be properly secured, and are not wobbly. You can also add an additional layer of protection by covering the tractor with a heavy tarp, netting, or roofing material to foil airborne predators like hawks. Double check that the covering is also securely fastened against wind.

You can install electric fencing away from your tractor to deter ground predators, or use motion-activated lights and sound devices. Certain plant and herb species like lavender, mint, and rosemary all have some deterrent properties and can offer some natural defenses. Some farmers use dogs, cats, geese, guinea fowl, llamas, and alpacas in and around the fields containing a chicken tractor, just be sure to choose the right animal for the job. A small dog, for example, may be a meal for a coyote rather than a protector, and remember that most guard animals require proper training

Keeping your chickens safe from predators will probably need a combination of approaches. Research and choose what works best for your situation.

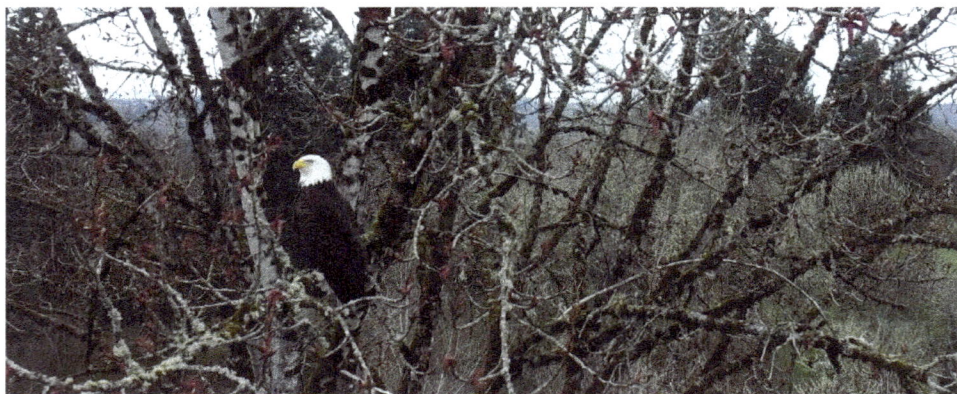

A Bald Eagle spotted in a tree above our fields trying to find an easy meal.

Danny's Details: We used a combination of methods, but the most notable and my highest recommendation is having an intelligent farm dog. They provide companionship, and an extra, better set of senses for the farm. Our dog Radar is an incredibly enthusiastic Aussie/border collie mix. He is wicked smart, and was not only trained on herding turkeys at a young age, but also instantly became aware of the hawks and aerial predators. When he was a puppy and would look up at birds overhead, I encouraged him and taught him the command, "Where's the bird?" That taught him to circle under the predators until they left. I still use this command today, and when I do, he will shoot out the door and scan the skies. He also fought off raccoons and chased coyotes off multiple times. His scarred nose evidence of his many battles.

One time he and our pug (imagine that power combo) trapped a great horned owl that was eating the heads of our ducks in our call duck pen, and racking up a heavy kill count. It was hilarious to see, and probably at least a little embarrassing to the poor owl who had a pug holding its wing tip through the fence, while Radar was barking and lunging at it through the gate. We let the owl go once I recovered the angry pug.

MAINTAINING A CLEAN AND HEALTHY ENVIRONMENT

Your chickens' health is important for their safety, for you, and your customers. A clean and healthy environment can prevent the spread of disease and parasites. Foot dips in shallow containers filled with water and disinfectant can be placed at the entrance to coop or pasture areas to help with this. Clean your coop or chicken tractor regularly, along with your water and food systems. Remove any debris and waste, and ensure any bedding is fresh.

Part of the reason for rotating your tractor to different areas of the pasture is to make sure they do not live in a build-up of waste underfoot. If a chicken becomes sick or injured, it's important to have a separate hospital or quarantine space to isolate it from the rest of the flock. This will help prevent the spread of potential diseases and allow the chicken to recover without being harassed by other chickens. If you do have an outbreak, thoroughly clean and disinfect your tractors and coops, and you may even want to consider rebuilding them from scratch. That may sound extreme, but with some diseases (including Avian Flu) that can be a legal requirement.

5

Feeding and Nutrition

Raising chickens that will become food requires you to have good food for them. That might sound obvious, but as a farmer, ensuring your costs align with your end product is important. Hopefully, this chapter will assist you in making smart and efficient decisions. You can give your chickens the most expensive gourmet food on the market, or stick with the cheapest, nutritionally lacking food around. Either path could be equally disastrous. Finding a good balance is key.

STARTER FEED

Starter feed is commonly given to newly hatched chicks for their first few weeks. Some of the specifics might change based on the manufacturer and the breed of the chicks, but most starter feed will contain the necessary components to encourage rapid growth and development. It will generally be composed of smaller particles for ease of consumption for chicks. Starter feeds should contain higher levels of protein (18-24%) to encourage muscle and tissue development.

Proteins can be plant or animal-based, depending on the manufacturer and desired mix. Plant-based proteins in starter feeds might include soybean meal, corn gluten meal, wheat, canola, sunflower, or cottonseed meal. Animal-based proteins range from fish meal, poultry by-products, and insect meal.

Soybean meal is considered a high-quality protein source and rich in essential amino acids. It is generally digested efficiently by chickens, resulting in good feed conversion and favorable growth rates. Due to its wide availability, it is also cost-effective.

Corn gluten meal is a moderate protein source and a byproduct of the corn milling process. It is not considered a complete protein source, and should be combined with other sources to create a complete balance. Given its lower protein content, corn meal may require a higher quantity, which could increase overall feed.

Wheat is another cereal grain-based starter feed that provides moderate protein content, but it is better known for its carbohydrate content. Wheat is priced competitively, and generally cheaper than soy or fish meal, but it can suffer from the same quantity vs. quality issues as corn meal.

If you're working with grains, you also need to contend with the risk of mycotoxins, which are often the result of improper handling and storage. Mycotoxins can harm chickens and lead to health issues and reduced performance. Ensuring strict quality control measures during storage and feed production is crucial to minimize the mycotoxin risk.

Fish meal is a high-quality protein source, rich in essential amino acids. As the name suggests it is derived from fish, and it provides a wide range of essential nutrients, vitamins, minerals, and omega-3 fatty acids. Mixing fish meal with plant-based proteins is a common practice to help counter the relatively high costs of fish meal. One potential drawback is that chicks may initially be reluctant to eat it, and it can take some effort to get them to acclimate to it. We actually tried a fish meal diet, but it made the eggs taste fishy and our meat quality seemed poor, so we stopped.

Poultry byproduct meal is a high-quality protein source from rendered poultry byproducts, like organs, bones, and meat. It is considered cost-effective and it's more palatable to chicks than some other options. The quality can be an issue, so you'll want to confirm how it is processed to ensure the meal is free from contaminants or other undesirable materials. Many folks cite an ethical issue with feeding chicken a chicken product, but in reality, chickens are carnivores. They always eat each other.

Insect meals such as mealworms or black soldier fly larvae are high-quality protein sources, and chickens generally find insect meal highly palatable. The major drawbacks are availability and cost compared to conventional sources.

Vitamins and minerals are crucial supplements often added to feed. This includes vitamins A, D, E, K, B-complex vitamins, calcium, phosphorus, zinc, selenium, and others. They contribute to bone formation, muscle development, immune function, enzyme activities, and other metabolic processes.

We used to offer poultry that were raised on a special diet, like corn or soy-free. We eventually stopped though. It was expensive to feed them this way, and typically the people who wanted this didn't want to pay higher prices. Once again, it was a simple determination of cost vs. benefit.

Another optional step is adding medications or additives to chicken feed. This can have several benefits, but it can also carry some detriments.

- Antibiotics can be added to fight bacterial infections. A nasty bacterial infection can spread quickly through larger flocks, but there is a chance that by adding antibiotics you are increasing antimicrobial resistance in bacteria.

- Nutrient supplements might be necessary if your feed lacks certain vitamins, minerals, and amino acids.

- Probiotics, prebiotics, and organic acids can be added to promote a healthy gut microbiota for your chickens.

- Growth promoters can be added to increase or maximize growth rates for better economic returns.

- Stress reduction additives can help alleviate stress by supporting immune systems and reducing the negative impacts of heat, transportation, or dietary changes.

The use of medications in feed could result in drug residues in poultry products, which can have implications for food safety and consumer health. Strict withdrawal periods and adherence to regulatory guidelines are essential to mitigate this risk.

GROWER FEED

Grower feed is designed for chickens in the intermediate growth stage, typically between 6-16 weeks of age, depending on the breed. It supports controlled development, muscle and skeletal growth, and overall health. It will contain slightly lower protein levels, balanced for growing your chickens at a healthier pace.

FINISHER FEED

Finisher feed is used in the final stages of your chicken's growth, and usually contains the lowest protein levels, and it can help you achieve the desired weight and muscle composition before processing. Finisher feed can impact meat quality, flavor, and texture. Some grains or meals can affect the flavor of meat, such as corn imparting a sweet taste, flaxseed meal with a nutty or buttery taste, and canola meal providing a mild flavor. This could be a good (or bad) way to distinguish yourself at the farmer's market. Once again, cost should come into consideration as you decide how to feed your chicks.

As discussed in earlier chapters, the composition of your pasture and the forage it contains will impact your chickens. Good insects for your pastures can be crickets, grasshoppers, mealworms, fly larvae, earthworms, and beetles. They can provide a good protein alternative, and encourage activity in your chickens.

Treats! While the primary diet should be a balanced commercial feed, it is perfectly fine to supplement their diet with treats like grains, greens, or other kitchen scraps. Be careful not to overindulge; too much of a good thing can cause issues.

Be careful of transitioning to different types of food too quickly. Do it gradually over a small period by mixing the old and new feeds. This should help transition to the new feed with the lowest risk of digestive issues.

FEEDING SCHEDULE

A feeding schedule is important, primarily so you don't underfeed your chickens. Underfed chickens will not only be underweight (which will ultimately cost you money), but could also suffer in different ways due to the stress it causes. Although there are exceptions and many different approaches to feeding depending on your end goals, we found a good rule of thumb to grow broilers in eight weeks is to provide a constant supply of food for your chickens. This is why it's important to have good feeders that provide ample access, and no competition for your birds. Keep them filled!

FOOD STORAGE

You'll want to keep your food stored securely to avoid it spoiling. Keep your chicken food in a cool, dry, and well-ventilated area, in airtight containers or bins. This should prevent moisture, mold growth, and pests. Buy only in quantities of food that will be used in a reasonable timeframe. Older food can lose effectiveness over time. Use a first-in and first-out method of consumption. Old food is not really a problem in the 1,000-bird method. Pests are the biggest issue. On our farm, when we were still experimenting with the amount of food to buy, it was the balance of how much we should get before the mice would get into it. We eventually found the right amount to buy each month, and we also started to store it in a trailer away from the barn.

Danny's Details: We used a feed that is milled about 100 miles away in Albany, Oregon. We had plenty of options locally, but we decided to regularly make the drive, since we got a better price at the hatchery (which also had a wholesale feed store). Each month, we took a flatbed trailer down and picked up one ton of feed, along with 100 chicks each month. The feed we used is a high-quality broiler crumble that we could use for all process phases. The chicks found it easy to consume, with less feed loss than others. We did occasionally use a starter feed, but the crumble is a great multi-phase grower, so we eventually just phased out using starter out of our system.

6

Processing - Slaughtering, Butchering and Packaging

There is some confusion regarding what is considered slaughtering and what is considered butchering. For the sake of clarity, we call the kill phase "slaughtering" and the cut-up phase "butchering." We distinguish between these things because they are done on different days, making it a more hygienic process as you slaughter in one kitchen, then chill and butcher in a freshly cleaned, or completely different kitchen.

The main reason we spread out the process is that fresh meat is less pliable, which makes it harder to cut. It takes time for the birds to get chilled and down to temperature, and we keep them overnight at temp. That lets the meat firm up enough to make butchering easier. Doing it this way also means the cuts will come out cleaner, which makes for a better package presentation to your customer.

HONORING THE ANIMAL'S LIFE

Killing animals for food is a deeply personal journey that anyone wishing to undertake should reflect on before buying their first chick. The reality is that you are taking potentially hundreds, even thousands of lives that you raised from babies. How do you come to terms with that? That's a personal question that only you can answer, but it's important that you do. The last thing you want is to find yourself with twenty birds ready to butcher, and you cannot continue. It is not fair to the birds, nor is it fair to you. One of the most important pieces of advice I can offer is this: before you start your own endeavors, participate in a butchering day or two.

If you find yourself questioning your resolve, just remember that there is a good reason for your actions. You are choosing to be a part of your community's food system. You are raising animals in a kinder, more sustainable way that will provide a healthier product for people. Seek out others who are raising birds and talk with them. Learn from their experiences and ask for guidance, advice, or reassurance. Raising your animals in an organic or free-range environment promotes a healthier relationship between humans and food animals.

I find comfort in taking a moment of silence and saying a few words to honor the animal. I realized my process on the first day, after the first animal I killed. But even before that, I had to confront this reality. A while back, there was a female turkey that would greet me every day. She was so sweet and would always jump up to say hello when I came in. I learned that these creatures trusted us completely. The truth is, I couldn't butcher that bird. Sharon took her away and handled the turkeys our first year. It forced me to confront the unmistakable reality of what was going on. I had hand-raised this turkey, and now she would die to feed my family. I vowed that from that day on, I would still love and hand-raise them, but I had to be the one to butcher them. I made that promise because they knew and trusted me, and I had to bring the circle to a close to honor their lives.

Coming from the city, many people only see a small piece of the process and don't understand the full complexities of the food process. Killing animals has become primarily an automated, clinical process that most people never see or participate in. There's a very real disconnect when our food animals are killed by machines. I'm not sure, but maybe that's something for you to consider as you read this book and consider your path forward.

My slaughter process is very focused on providing a calm experience. The animals can sense things; if you come in nervous, scared, or sad, they can sense it. On slaughter day, I take time to calm my thoughts and emotions, while focusing on everything I have to be thankful for. My life, my farm, my children, the way we live. I go over all of this in my head before undertaking the day's efforts. My wife then does a short blessing for us and the animals. When I begin, I visualize the perfect kill.

In the final moments of their lives, I hold the birds. They know me as the one who fed, watered, and talked to them every day. Then I walk them over to the kill cones, turn them upside down, and place them in the cone. I massage their neck to calm them down, move the feathers out of the way, place the extra sharp knife on their neck just above their cheekbone, take one deep breath, and say, "Thank you for your life," as I draw the knife quickly. It is important to make the cut just deep enough to sever the artery, but not so deep that it hits the spine or esophagus areas where many nerves are. I do this 1,000 times a year. If you don't take time to honor each life individually, you begin to take those lives for granted. I don't believe that should ever be done.

Take pains to honor the lives of the chickens, and never approach this in a nonchalant way. We strongly encourage you to empathize with your animals and commit to an ethical and efficient butchering process.

Slaughtering Tools

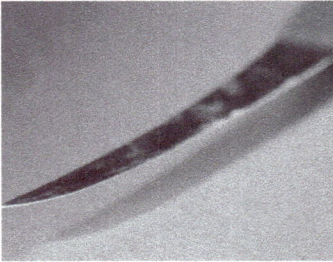

Knives

- A **sharp and sturdy knife** is essential. It should have a well-honed blade to quickly and effectively sever the bird's arteries in its neck.
- A **dull pinning knife** for removing any stray feathers that may remain on the bird.
- A sharp, **small-bladed knife** for opening up the chicken's body cavity and removing internal organs.
- A **boning knife** and **poultry shears** may be useful for cutting and separating the head and feet.

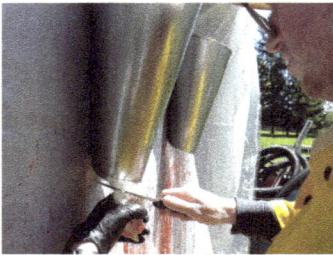

Killing Cones

Killing cones are metal or plastic cones that hold the chicken upside down to facilitate humane and efficient slaughter. They restrict the bird and are hung from a rack with the narrow end pointed toward a metal or plastic receptacle designed to catch the blood flowing out of the animal. The bird's head is pulled through the lower hole to offer easy access to its head and neck.

Scalder

A scalder is a heated water container used to submerge and scald the birds after slaughter. This helps loosen feathers and makes them easier to pluck. A large pot or tub with temperature control can be a scalder for small-scale operations. The temperature should be heated to 148°-152°F (63°C).

Plucker

A "plucker," also known as a plucking or picking machine, is a device that removes feathers from the bird's body. It consists of rubber or plastic fingers that rotate rapidly and remove the feathers from the chicken as they bounce around inside the machine.

Sturdy and Clean Cutting Surface

A sturdy and clean cutting surface, such as a butcher block or a stainless-steel table, is necessary for cutting the birds. Ensure it is sanitized properly before and after use.

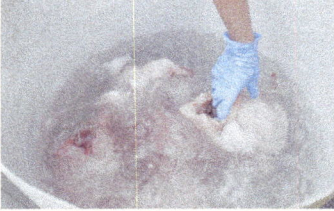

Chilling Tank

A chilling tank capable of holding and fully submerging all of the animals you will be processing. The most common way to do this on a small scale is with a cold-water bath at 40°F (4°C), achieved by filling a tank with ice and water. The temperature of the tank should be regularly monitored.

Containers

You'll need separate containers, also on ice, to collect and store the different parts of the chicken, such as giblets, feathers, blood, and offal while you go through the processing phase in order to keep food safe.

Hoses

Hoses and spray attachments for pressurized water.

Safety Gear

Safety equipment, such as goggles, gloves, cutting gloves, and aprons.

Non-stick Cooking Spray

A non-stick cooking spray to be used on your equipment (especially kill cones) before beginning the butchering process will make clean-up easier.

Food Grade Unscented Soap

Add to the scalder to help penetrate the birds feathers down to the follicles. This assists in feather removal.

SLAUGHTERING

To begin the slaughtering process, you should stop feeding your chickens 24 hours prior, so that their digestive systems have a chance to empty. This can help make the process more hygienic as the gizzard will be empty, but it can also help keep the birds calm since their energy levels will be lower. Move your finishing tractors to within easy reach of your processing kitchen, and close to the kill cone station.

Slaughter

Take the individual birds and place them head down within the kill cones, with their head coming out of the bottom hole. Occasionally a chicken may struggle, but the cones will hold them securely, keeping their wings close. Once you have the chicken in the cone, pull gently down on the head, and with your kill knife, **sever the carotid artery just under the cheek and ear hole on the side of the neck.**

Be careful not to go too deep, and avoid severing the windpipe, esophagus, or spinal cord. Another reason to cut as high on the neck as possible is that when you remove the head later, you'll want to separate as high on the vertebra as possible, so that you don't lose neck meat and retain sell weight. Let the birds sit in the kill cones for a minute or two and allow for proper exsanguination before removing them. I tend to pull them out of the cone and check to see if their wings are limp and away from their sides to ensure they are fully bled out, and their nerve twitching is complete. Once that is done, I put them in the scalder.

On a side note, simply removing the heads right away is possible, but it can potentially put your birds into shock. That could trigger adrenaline or cortisol, which can change the flavor of the meat. It can also stop the heart from pumping, which would limit the amount of blood that' pushed out of the bird.

Scalding

The next step in the process will be scalding and feather removal. Your scalder should be set up next to the kill cones, and the water temperature should be at 148°-152°F (63°C). If the water is too hot, it can cause the skin to tear or cook the bird, making feather removal more difficult; if the water is too cool, it may not effectively loosen the feathers. A small amount of food-grade dish soap, vinegar, or citric acid can be added to the water for easier penetration of the feather follicles. Your scalder can range from a large pot over a fire to a high-end commercial-grade machine. The important part is that the birds are fully submerged and constantly agitated through dipping, or a rotary dunking system, and they do not start to cook in the water. It's worth mentioning that while dipping them by the feet works, the feet tend to need the most scalding, so a rotary dunking system is preferable.

Let your birds go through the scalder for 60-90 seconds, adjusting things as needed based on the size and age of the birds.

Plucking

Once you take the bird out of the scalder, you can remove the feathers by hand, or by using a plucking or picker machine (recommended). Using the machine method, have a spray hose with cold water ready, and douse the birds as they rotate. The cold-water spray helps make the feather removal faster and easier, and helps wash the chickens off. Most pluckers have cold water hose attachments that spray as the birds spin.

Removing Head and Feet

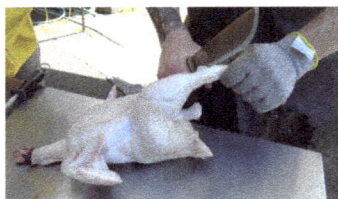

Once you are finished removing the feathers, move the chickens over to your cutting surface. Do a once-over of the chicken and clean it of any manure or other filth. Take off any stray feather pinions remaining as well. Remove the head at the closest vertebra to the head, either by cutting or pulling. Remove the feet by cutting at the lower joint under the drumstick and between the bones. If you hit bone, you are not doing it right. To find the correct place to cut, bend the joints in the opposite direction from the way they naturally bend, then look for the valley that develops. That will show you where to cut. Cut a slight nick to make the valley more pronounced, and then continue cutting where the V develops.

Cleaning

Oil gland removal is optional

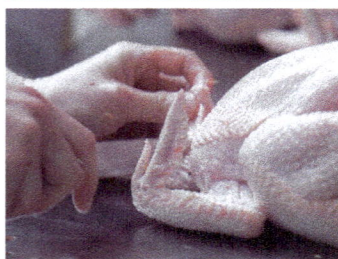

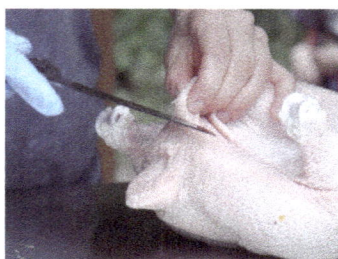

Flip the bird onto its belly, and at the top base of the tail there is an oil sack you will want to remove. At this point make sure the bird is on it's belly with it's tail pointed away. Grab the tail with your thumb and forefinger in a pinching motion to hold the oil sack tight while you cut it off. Cut down about a quarter inch at the base, then directly outward toward the end of the tail nub. Move the bird onto its back and find the spot at the top of the breast and just beneath the neck. Pinch the skin in the valley the breast creates, and pull up before cutting a small hole enough for your finger to go through. Tear the skin open and find the crop at the top of the breast meat. It will appear as a fleshy sack a few inches wide. If you stopped feeding your birds 24 hours earlier, it should be empty. It's slick and hard to grab, but you'll want to take the time to remove it by picking it away without damaging the breast. Once complete, you'll be able to pull the windpipe and esophagus out of the neck. Do not cut these off yet, but let them sit after loosening them from the breast and neck. The goal is to avoid rupturing or cutting any part of the digestive tract.

Turn the bird so its tail is pointed toward you and the breast upwards, and cut the loose skin above the tail and vent (between the vent and lower breast). Try to cut closer to the vent to save skin and fat, then pull open the body cavity. Again, be sure not to sever any of the digestive tract. Reach in and pull out all of the guts and offal. Pay special attention to hooking the esophagus, pulling both it and the crop out of the bird without any breaks as you pull the rest of the internal organs out. Lay the organ pile on the table to remove the gizzard, heart and liver.

To save the liver and heart you must first remove the gall bladder. On the liver you will see a inch long slug shaped organ, it will be the darkest organ you see, a kind of dark greenish black. Hand pinch off the gall bladder from the liver. You don't want it to split or break, as the liquid contained within is very bitter and undesirable on your meat. BE CAREFUL!

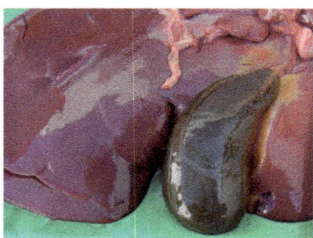

Lucyin, CC BY-SA 3.0 via Wikimedia Commons

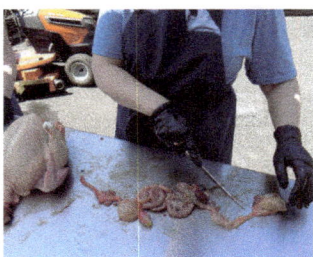

Once the giblets or offal are removed and on ice, pull the remaining viscera out taut, stretching the intestine outward from the vent. Cut down on either side of the vent and around, separating the intestine and remove. Finally, reach back into the body cavity and remove the lungs by scooping the side walls along the spine with your fingers in a scooping motion, the lungs will be a bright pinkish red. Take your spray hose and clean the body cavity, and do a final exterior wash before depositing your bird in your chill tanks.

Chilling

The chill tank is a relatively straightforward system and process to get birds down to food safe temperatures. A food safe container filled with ice and water should be cold enough to drop the internal temperature of the bird to about 40 degrees within two hours. On hot days be prepared to have lots of extra ice, you will probably want to buy an ice machine for cost savings. Make sure you monitor the birds and record how fast they drop in temp this can be required documentation to keep on hand for inspections. Make sure you know your local food safety guidelines.

Alternately, you can invest in an air-chilling system, much like modern HVAC Systems in function and price, which may be prohibitive to small farms.

Organ Disposal

Make sure you are prepared to deal with the entrails. The last thing you want is to forget about the gut bucket for a few days in the middle of summer while you are off selling your product. Some areas allow you to bury them, but others you will require you to take it all to the dump.

BUTCHERING AND PACKAGING

By now your birds should have rested for 24 hours in the iced down coolers or refrigerator and were cooled down to the food safe temperature in the required time, see your local regulations. In Washington state the bird must be to temp within two hours of kill/being put in ice and the temperature must be recorded and available for inspection.

At this point, the meat has a good firmness, and will be easier to cut and package. Pat your chicken dry with paper towels to make it easier to work with.

I will run you through the basic cuts. You will be dividing the bird into leg quarters (leg and thigh together), wings, breasts, tenders, and carcass for your prime cuts.

Butchering & Packaging Tools

- Knives and poultry shears
- Cut gloves
- Cutting boards
- Offal and parts bowls on ice
- Vacuum sealer and bags
- Large pots
- Heat shrink bags for full sized birds
- Sharpies, zip ties, labels
- Scale

- Paper towels, soap, and bleach
- A refrigerator or freezer for storing the butchered chicken. The temperature should be kept at appropriate levels for food safety. (Be aware that depending on the type of permit you acquire, you may be unable to freeze the birds you wish to sell to the public)

Leg Quarters

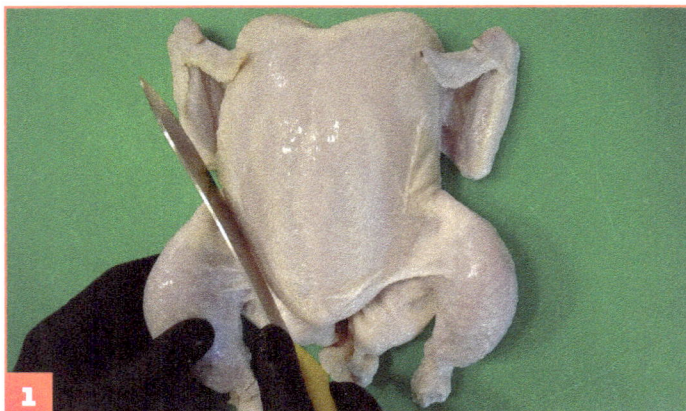

Slice through the skin just below each side of the breast, between the breast and legs.

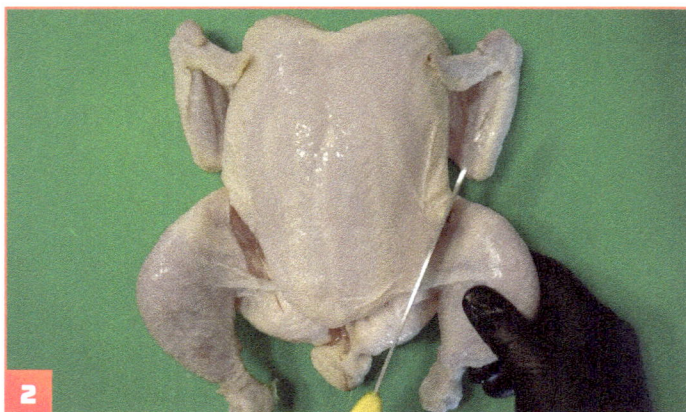

Slice the fleshy section of skin there that is not attached to any muscle. Slicing here allows you to manipulate the legs out of the socket for easier separation.

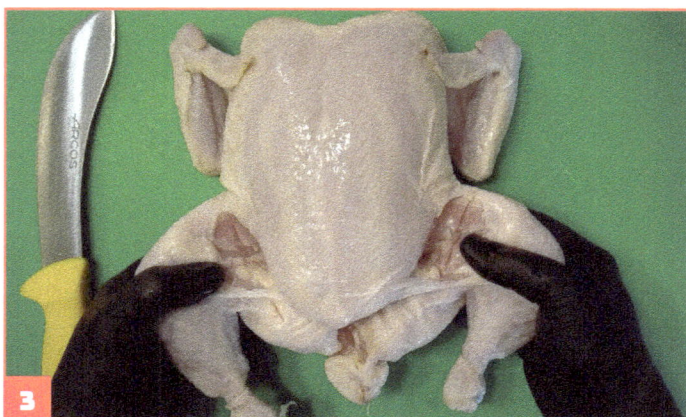

Lay the chicken breast up, grab both the legs with thumb and forefinger, and put the rest of your fingers behind the bird, supporting the spine.

Leg Quarters

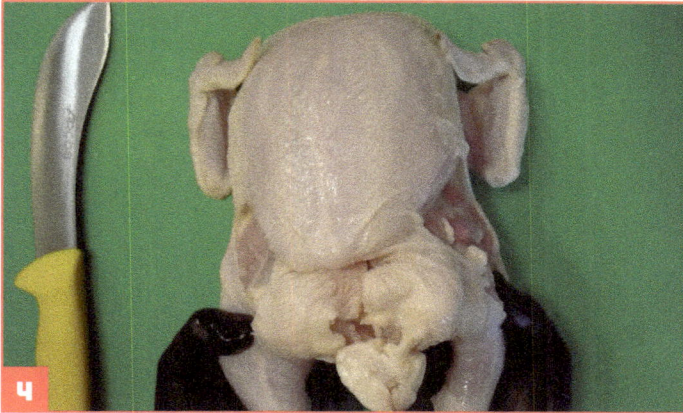

While holding the legs and supporting the back, rotate your thumbs outward, pushing the legs out and back, and push the spine up toward you at the same time, this dislocates the thigh from the carcass making for an easier cut.

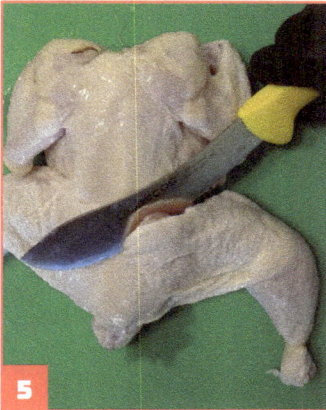

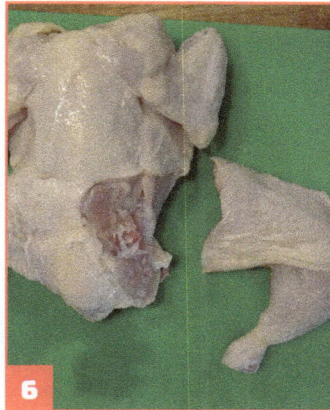

Take your knife and draw cut along the top of the thigh, flip the bird as you go, and go around the dislocated thigh bone.

Then finalize the draw cut down, and toward the vent. You should now have a leg and thigh quarter.

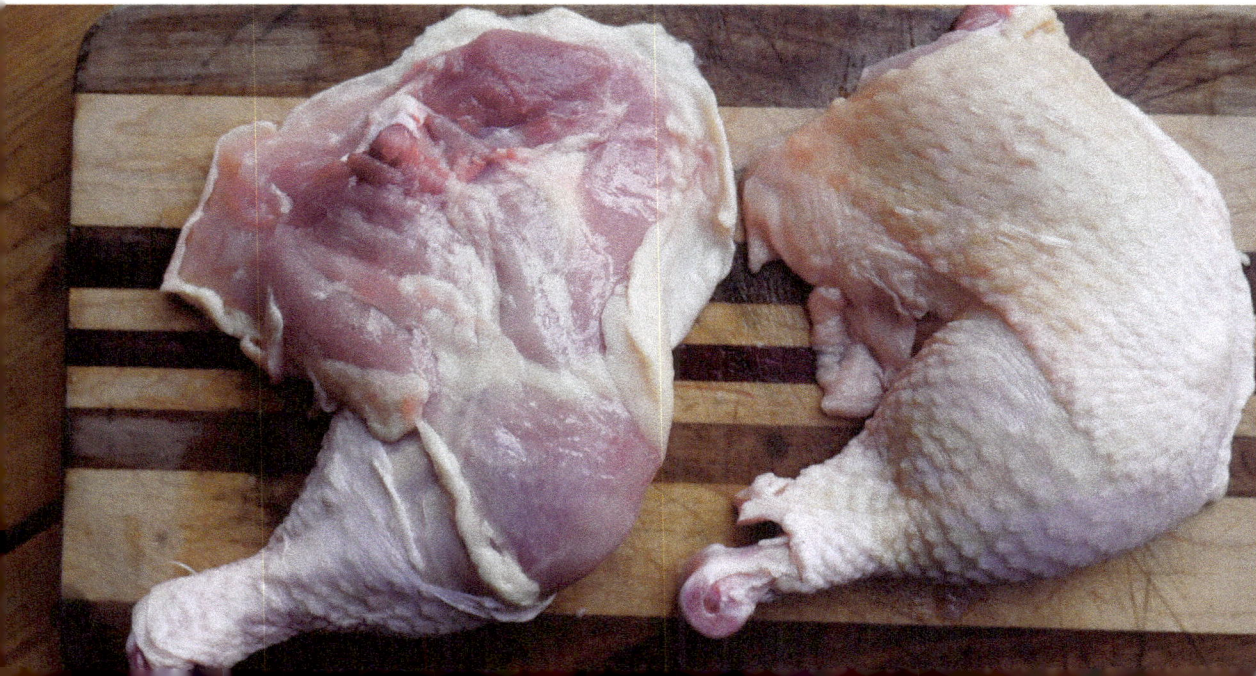

Specialty Cuts: Drumsticks

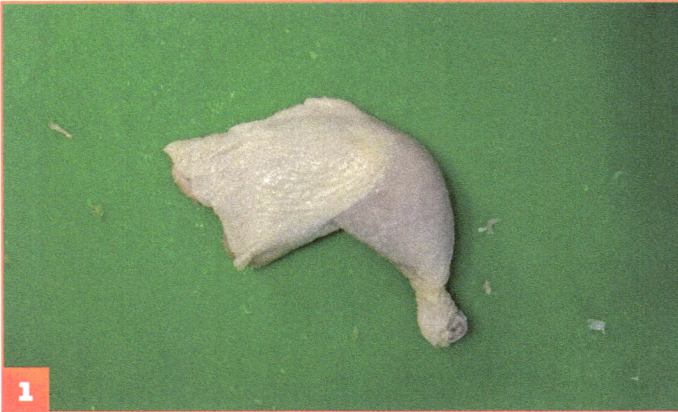

You can further separate the thigh meat from the drumstick by cutting between the remaining joint and along the fat line.

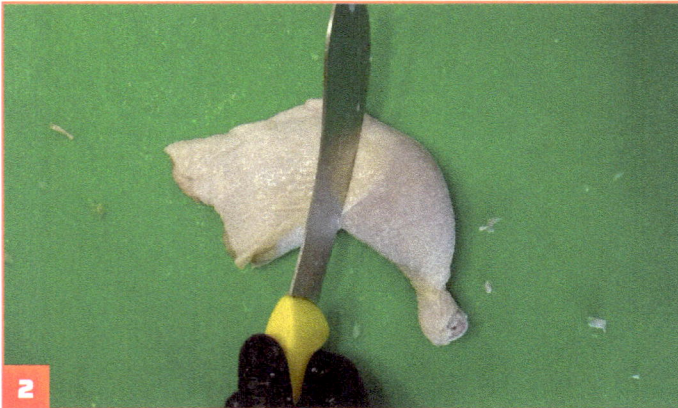

The blade will need to cross the joint at a 45-degree angle and along the fat line.

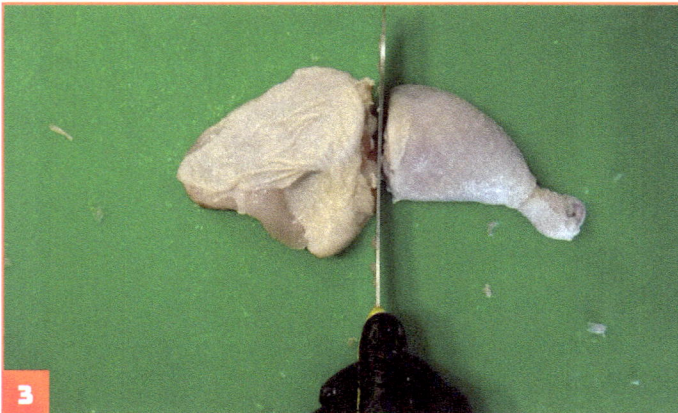

With a strong push and draw cut, you will get through the joint.

Specialty Cuts: Boneless Thighs

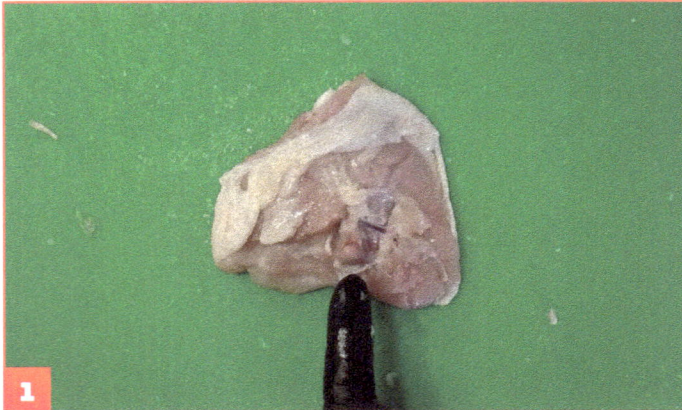

For boneless thighs, find the two bony socket ends.

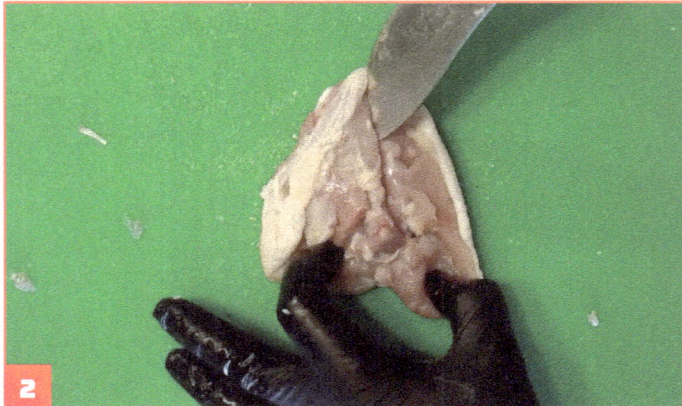

Using the tip of your knife, cut down to the bone and connect the ends with a smooth straight cut.

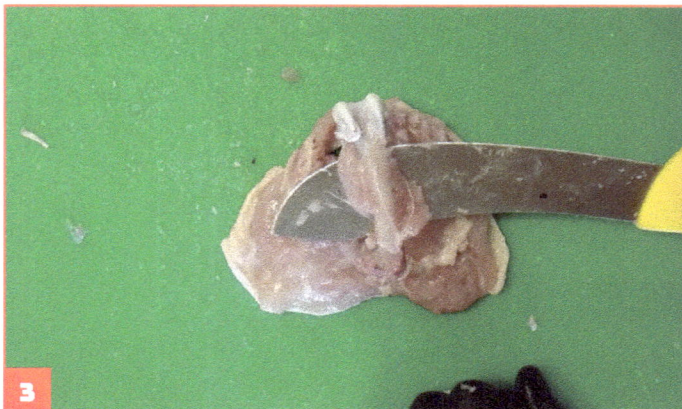

Work your knife tip around the bone, and gently pull to remove the bone.

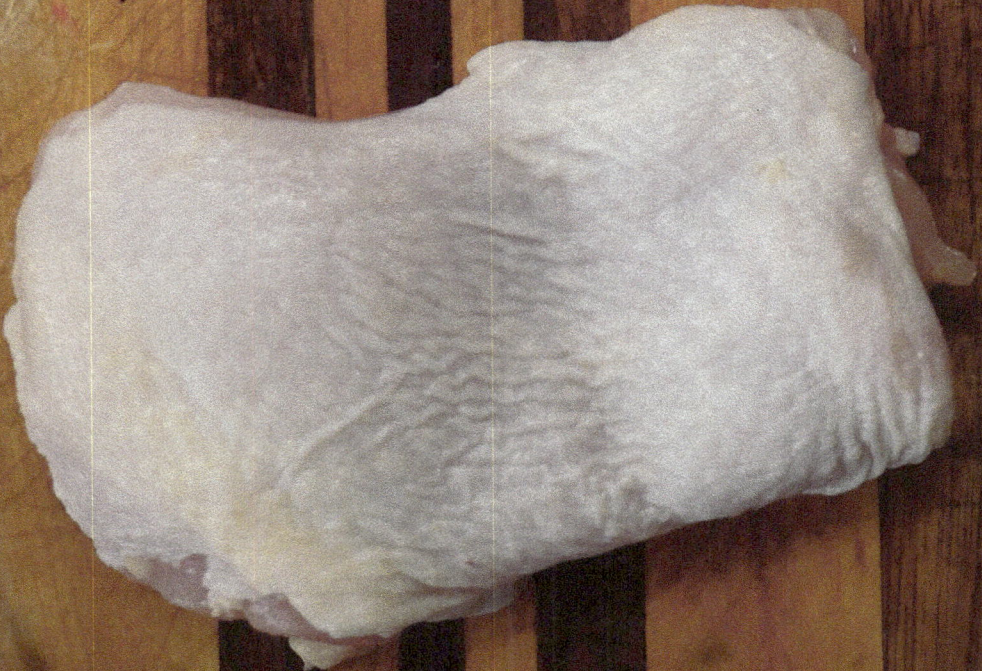

Wings

Take your knife in one hand and manipulate the wing with the other, and start on the back at the shoulder blade.

Cut in a smooth circle around the wing and toward the breast, as close to the bone and joint as possible, to separate connective tissue.

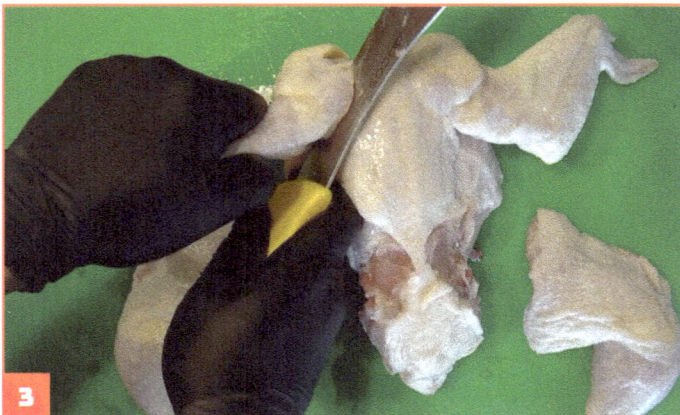

Be very careful not to damage the breast in this cut. Repeat with the second wing.

Specialty Cuts: Wingettes (Party Wings)

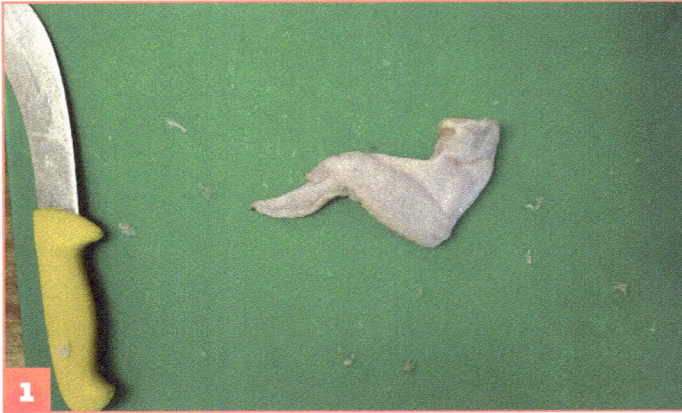

Cut the wing tip off at the joint, (image 2) then position the knife between the upper drumette and lower wing.(image 3) With firm pressure and at a 45-degree angle, cut down and draw the blade through the joint.

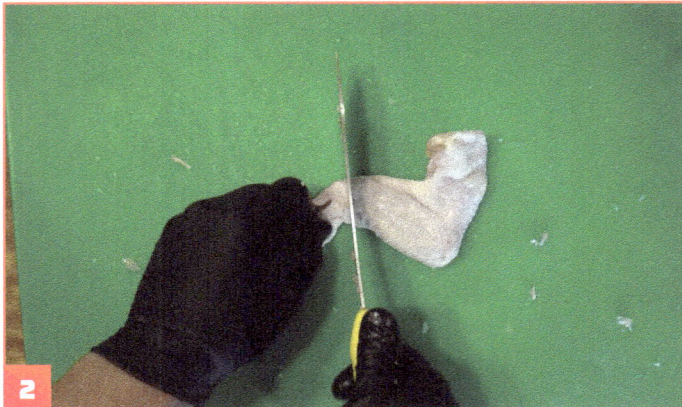

Breasts

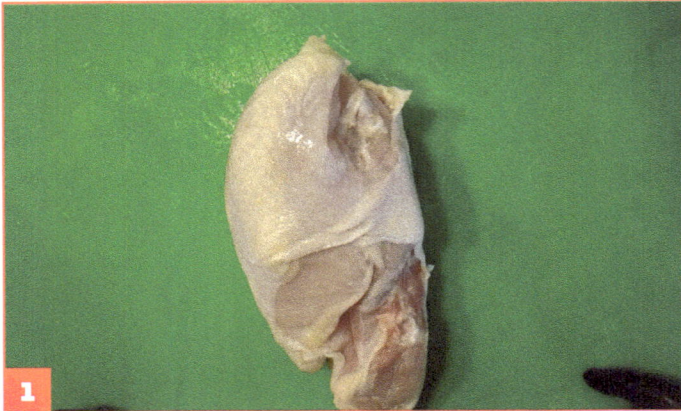

To remove the chicken breast, lay the carcass on it's side.

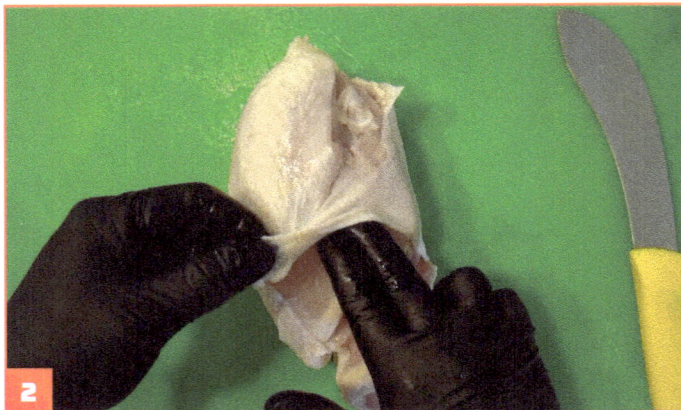

Peel back the skin by sliding your fingers up between the skin and breast and peel the skin gently away from the breast, until the breast meat is exposed.

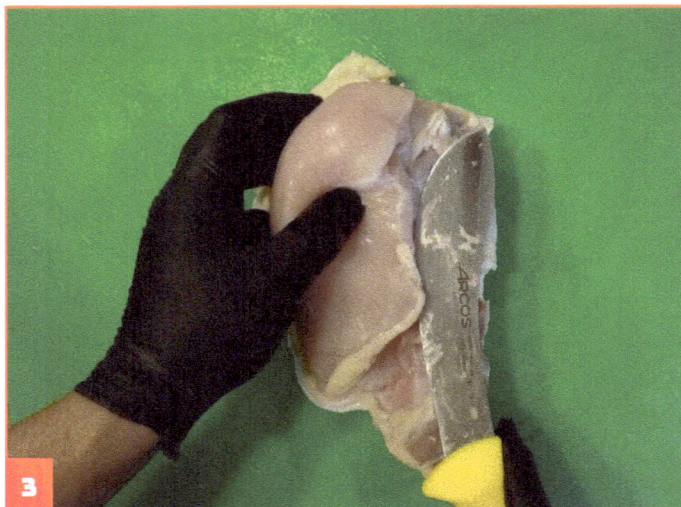

You will see a fat line that connects the skin to the rib cage closer to the spine. Take your knife and lay it gently on the fat line. Do a gentle pulling stroke along the line with the blade at a very shallow angle, to break the breast meat away from the back and rib cage, while simultaneously pulling the breast toward the chest keel to aid in tension to create a cleaner cut.

Breasts

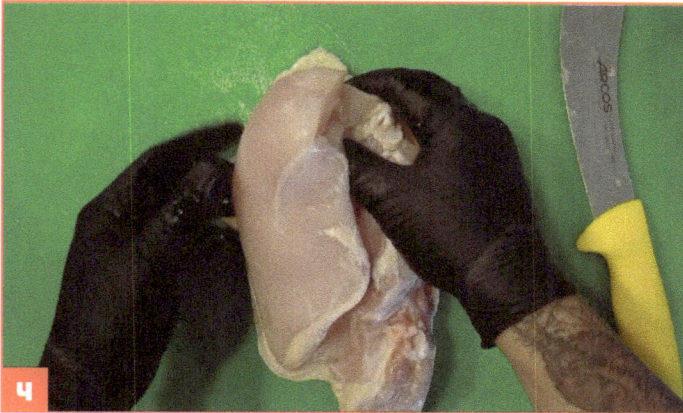

Put your knife aside and slide your thumb into the cut between the breast and the rib cage.

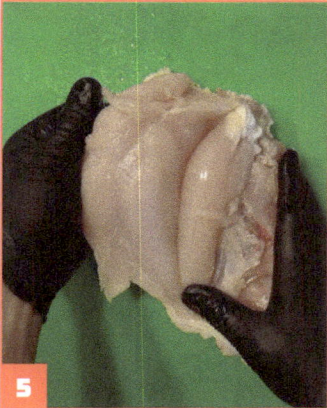

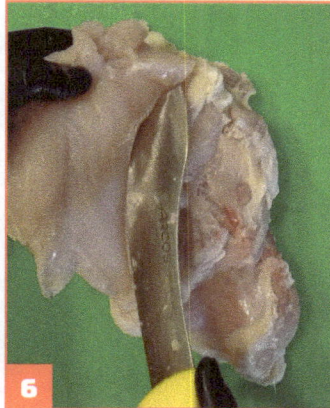

Work your way toward the keel bone moving back and forth until you have the breast separated from the breast tender underneath; this is more like tearing it away, but it is a better overall technique.

The breast should be connected at the keel only now. Use your knife and make a clean final cut on each side to make breast halves.

Specialty Cuts: Whole Butterfly Breasts

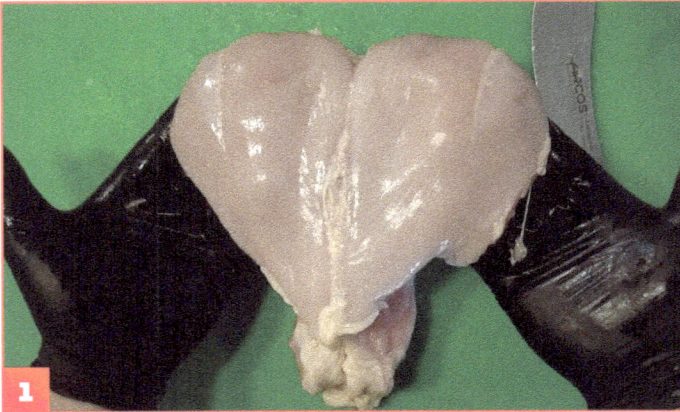

If you choose to make a whole butterflied breast, do not cut off the halves.

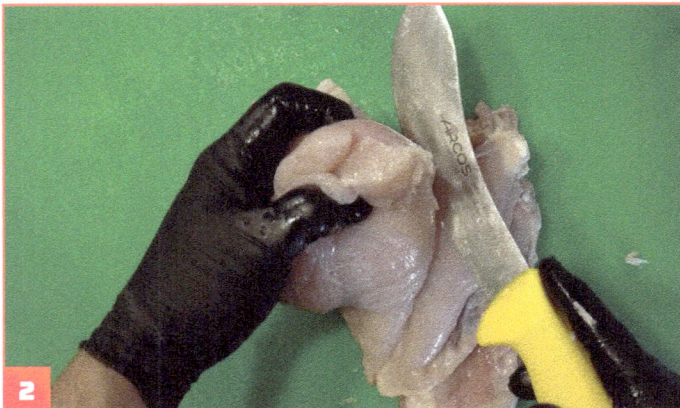

Instead, work your knife as closely to the keel as possible on both sides until you get to the top of the keel. Grab both breast halves in one hand, then take your knife and cut along the top of the keel.

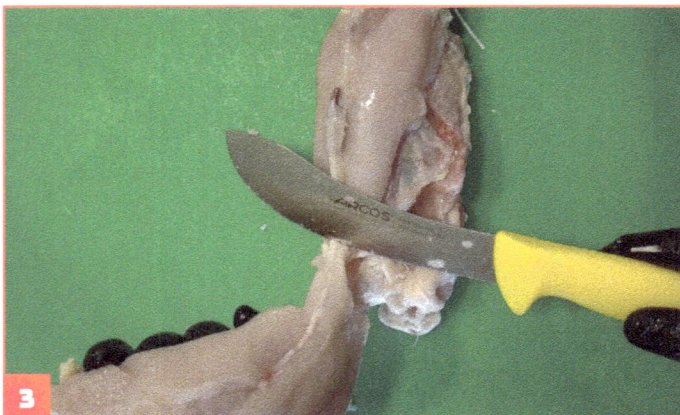

The whole breast will be intact and the carcass will fall away.

What you have is a full boneless breast still connected and shaped like a butterfly.

Tenders

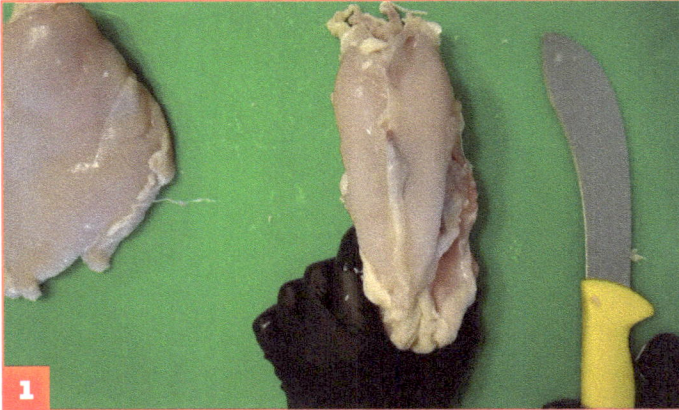

Lay the carcass on its spine with the keel bone pointed up toward you. Start your cut at the top of the wishbone (choose either side) and cut down and across to the opposite side keel bone, then do the same on the other side.

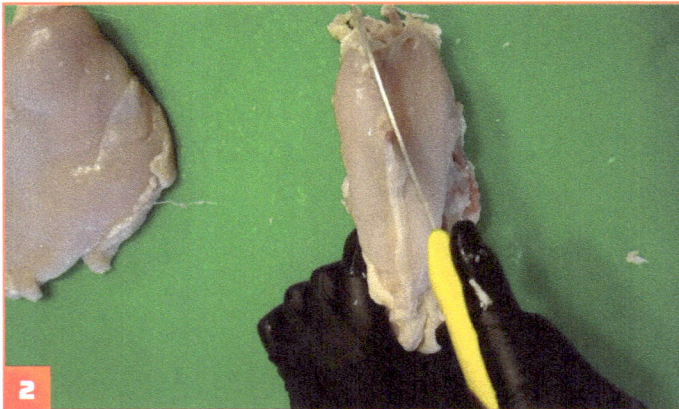

Essentially, you are making two cuts to form an "X," from the top of the wishbone to the bottom of the keel on either side.

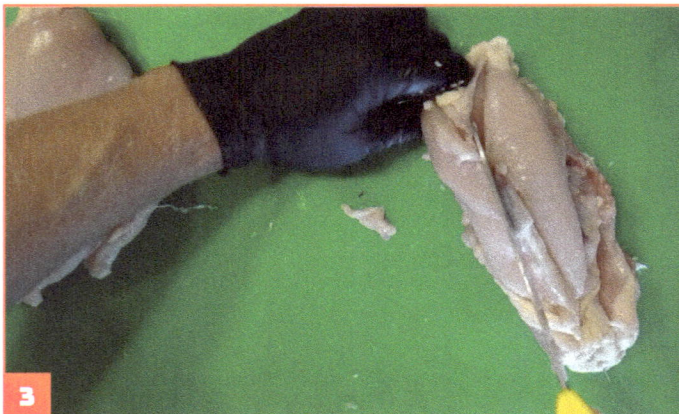

Once you've completed the cuts, run your knife under the tender and free it from the back and keel bone, you should now have two chicken tenders.

Carcass

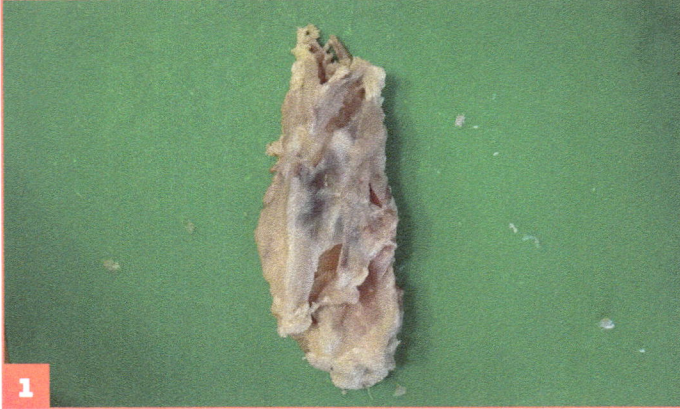

After the tender cuts are complete, you should have a carcass with 95% of the meat removed, including the best choice cuts. The carcass, or what we called the meaty back, is all that's left!

Specialty Cuts: Backstraps

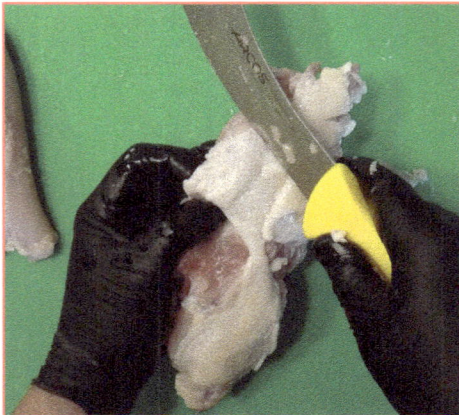

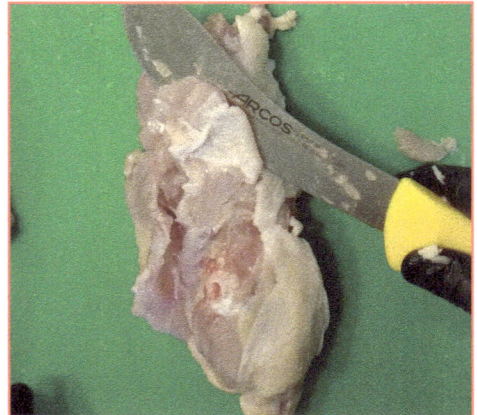

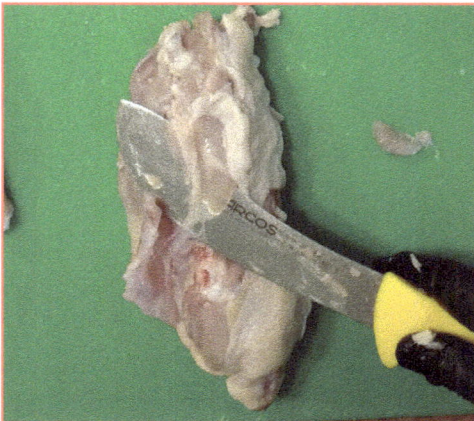

On the carcass back, there are two small nugget sized cuts called backstraps (or back tenderloin on other animals).

After a butchery session I typically hoard them for the family rather than sell them. They are so good that you can sell them for breast weight pricing. But if you are making the effort, I'd recommend keeping them for yourself!

Prime Cuts

1. Breast Halves
2. Full Wings
3. Tenders
4. Leg Quarters

Specialty Cuts

1. Butterflied Breast
2. Wingettes
3. Drumettes
4. Tenders
5. Drumsticks
6. Boneless Thighs

Once you have completed the butchery, you can package and freeze the chicken parts (including the neck, liver, and heart) for your family or prepare them for sale. If you can operate under a license that allows for parting out and freezing to sell to the general public, you'll want to cool the whole carcass down to 40°F (4°C), then let the meat rest for 12-24 hours before parting and freezing. While it is possible to chill, then part and freeze right away, the meat will be less tender than if you let it rest.

Specialty cuts are a whole different ball game in terms of how you can maximize your profitability. There are all kinds of cuts that can be done in butchery, and if you are interested, there are some very good books that specialize on this topic alone. If you take the time to become proficient with these specialties, you need to charge for them. It ultimately comes back to whether or not it's cost effective. A great example of the time vs. market value debate is the gizzard. The gizzard is a problematic cut for a lot of reasons: it takes a bunch of time to properly butcher, it is a messy process, and it can contaminate the other cuts and surfaces. So, can you sell them at a price that makes it worth your time?

You have to remove the gizzard while you are processing the bird. You'll need to put it in a separate iced area, then process it on its own by cutting it open and peeling the hardened inner lining away from the muscle, then dumping all the insides. The gizzard is part of the chicken's stomach process, and where it grinds its food. Chickens eat rocks and grit and store them in the gizzard, which grinds grass, grain, and bugs down before entering the chickens' digestive system.

Of course, the easiest thing to do with the gizzard is to just throw it away with the rest of the innards. We don't save the intestines, lungs, gall bladder, etc., as they pose too much contamination risk for what we do sell. Still, you may find some uses for these organs, and there is a market for things like gizzards. You have to balance the time it takes with the demand and the price people are willing to pay. There is a whole list of things people will eat, including things like the feet, combs and waddles. The more buyers you can find for this, the more it will make sense to take the extra time.

For us, there are a few specialty cuts I like to do; one was a whole butterflied breast, the other was the back strap nuggets. The party wings and thigh drumstick separation are also good, but I found that keeping the leg quarters together in our market was the best choice. You will need to experiment with selling cuts, and get your customers involved. We started doing party wings because a lot of people don't like wing tips.

Packaging is another thing entirely, and can be a vital component to selling your birds. I highly recommend purchasing a commercial-style vacuum sealer (unless you enjoy screaming at inanimate objects). We employed two different methods for our packaging: vacuum sealing, and hot bath heat shrinking. We like doing the heat-shrink-style bags for whole birds. You quickly dunk the packaged bird into a 180-190°F degree bath, and the plastic heat shrink, shrinks, and makes a nice-looking end product. When vacuum sealing, take the extra time to position the parts you are selling in a way that is appealing to look at, and will freeze nicely. You want your customers to be excited to use the high-quality meat they paid for. Just throwing it a bag shows a level of disdain that your client base may take issue with. Take the time to make things look right, and treat the meat carefully. It matters to your customer.

We separate our parts into a few different packages. First off, the highest priced meat is the breast and tender. This package alone can pay for the whole bird if you do it right and the demand is high enough. Use the cuts I discussed above (specifically the butterfly breast cut). Lay them out nicely with the best-looking side visible through the smooth plastic part of the vacuum bag, not the wrinkly side.

Package wings in eight packs, party wings in sixteen packs, leg quarters as two, thighs as four, tenders as eight, legs as four. You'll want to make stew packs or puppy packs (treats for pets) for all those extra chunks that don't make the cut. Feet are hot items for those that like bone broth. If you have the right type of processing kitchen, you can easily get the approval to make frozen broths and herb packs to go with your sales. Depending on who frequents your booth, you may want to do smaller bags. But keeping your price points at around $20 can help with sales, since that's what comes out of an ATM. That may not matter as much these days thanks to card sales, but I always want to make it easy for my customer to spend their money.

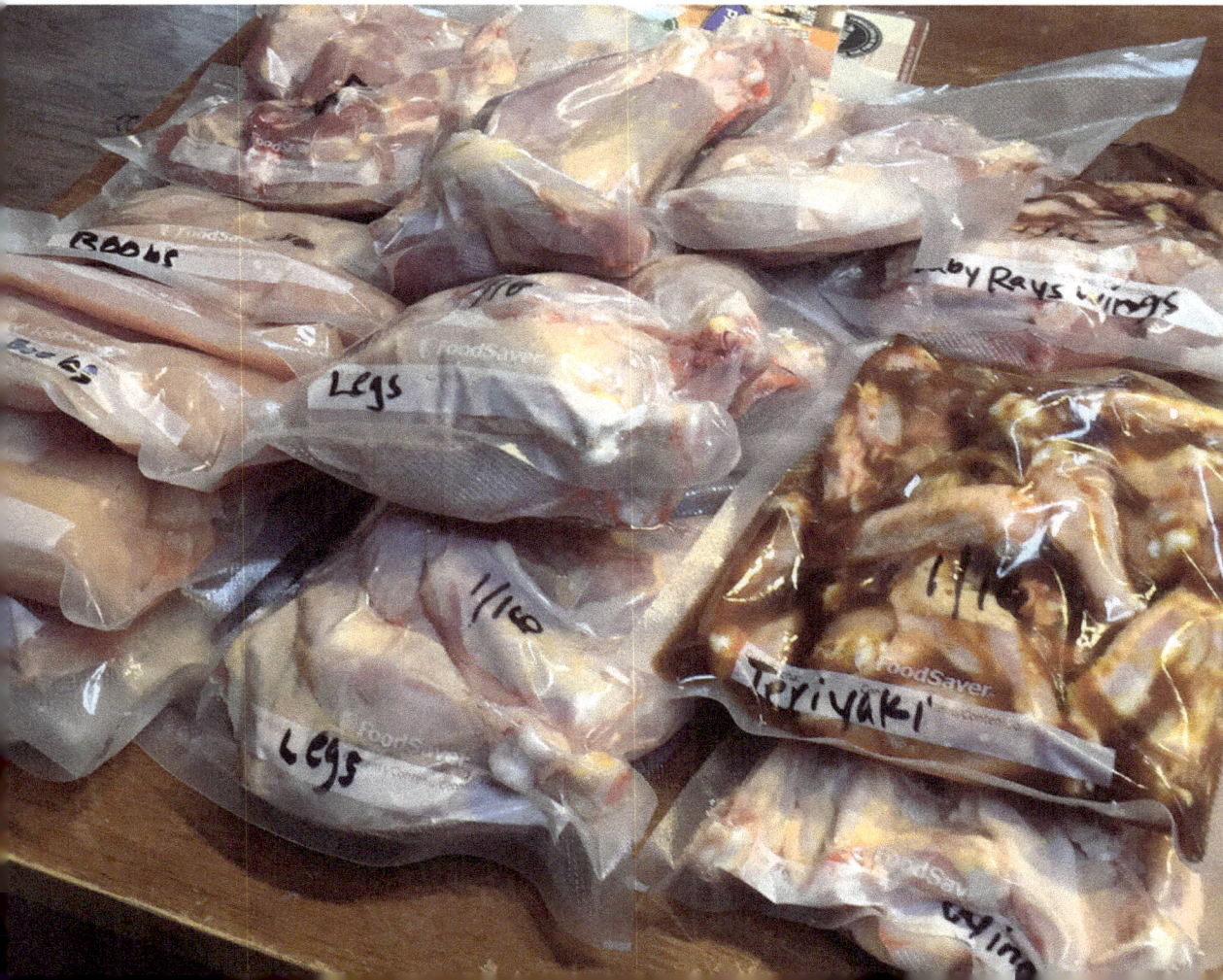

7

Certification, Licensing, and Insurance

Different licenses allow you to process your chickens in different ways. There are several ways to get licensed to sell poultry in Washington State, where we live, but you'll want to do your research and adhere to local regulations. The permits listed below are specific to our region, but other states will have similar permits, so consider this a guideline of the type of issues you may face.

WSDA SPECIAL POULTRY PERMIT

The first permit level is a Washington State Department of Agriculture (WSDA) Special Poultry Permit. This allows you to raise and sell up to 1,000 birds per year. It also stipulates that you may only sell whole birds within 48 hours of slaughter, and they must be chilled to 41°F (5°C). You may not freeze, part out, vacuum seal, or further process the birds. You cannot sell directly to restaurants, grocery stores, or farmers markets, and you can't ship your slaughtered birds via the mail or other services. You can only sell directly from your farm or slaughter location.

To acquire this permit in Washington state, you can apply to the WSDA Food Safety Program through the Washington State Department of Agriculture. Make sure you apply at least six weeks before your first scheduled slaughter. Along with the application and fee, you will need to provide the following:

- Poultry slaughter/preparation site diagram.

- Detailed processing steps and/or a flow diagram. Include as much detail as possible for each step (e.g., kill, scald, pluck, eviscerate, rinse, and chill).

- Water supply testing results.

Once the WSDA (or the organization from whatever state you are in) has received the application, a food and safety inspector will come to your farm and verify that the facility is adequate and the slaughter and processing are done in compliance with state food safety requirements. Inspections include an evaluation of personnel, the grounds, butcher facility construction and design, sanitary operations, pest control, sanitary facilities and controls, equipment and utensils, processes and controls, labeling, and licensing. Preventing overhead contamination, having food-grade surfaces, and chilling are key areas of concern.

Schedule your inspection on a day when you can do a dry run. The inspector will tell you if you have passed or failed, and further steps may be required. Once approved, your certificate will be mailed to you.

WSDA FOOD PROCESSOR LICENSE

Another permit is the WSDA Food Processor License. While we will be talking about this license in relation to poultry, it also covers many other forms of processed food, including dried fruits, jams, salsa, sauces, dried herbs and teas (other than just air-dried), bread, cookies, cider, and post-harvest mixed salad greens It also covers seafood, dairy products, baked goods, canned products, and condiments. This license allows you to slaughter up to 20,000 birds annually, process and break down birds into parts (parted out), and vacuum seal your frozen birds. You can also sell your product from the farm, farmers markets, on the internet, to hotels, restaurants, food service institutions, grocery stores, and via wholesale food distribution in Washington state. You may not sell outside of your state.

Once you apply for this license, Food Safety staff can offer one-on-one technical assistance with the licensing process. The application itself will include sections on the following:

- Sanitation schedule.

- Detailed processing steps or flow diagram. Include as much detail as possible concerning the steps involved (e.g., kill, scald, pluck, eviscerate, rinse, and chill).

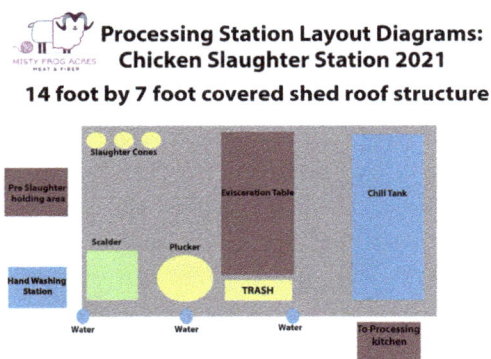

Processing Station Layout Diagrams: Chicken Slaughter Station 2021

14 foot by 7 foot covered shed roof structure

- Poultry slaughter/preparation site diagram.

- Proposed labeling.

- Water supply and testing (you'll want to allow plenty of time to approve your facility's water system, which can take weeks or months).

Once the Food Safety Program has received the application, a WSDA food and safety inspector will come to your farm and verify that the facility, slaughter, and processing are done in compliance with state requirements. Facilities, lighting, bathrooms, and water (from private wells) requirements will differ from those of the Special Poultry Permit, although the sanitation, cooling, and refrigeration requirements will be the same. Unlike the Special Poultry Permit, the Food Processor License will require a separate facility dedicated to the commercial processing operation. Slaughter and plucking may occur outdoors, but the rest of the process must occur indoors. Unannounced inspections may occur every 6-36 months.

After 20,000 birds, it becomes a USDA FSIS-inspected facility, which is beyond the scope of this book.

FOOD PROCESSING FACILITY REQUIREMENTS

Processing equipment should be made from easily cleaned materials, and be in good repair. Stoves, refrigerators, dishwashers, and other appliances or motorized equipment do not need to be commercial grade, and should similarly be made from easily cleaned materials and in good repair.

Worktables and counters should be in good repair, clean, and non-corrosive. Stainless Steel and high-impact, scratch-resistant plastic are most commonly recommended. Metal, finished wood, or hardwood used for cooking are also acceptable. Unfinished wood is not acceptable. Floor materials should be durable, readily cleanable, and in good repair. Sealed concrete, vinyl linoleum, or tile flooring is good, unfinished wood or other porous material is not suitable. Walls must be covered with a washable, non-porous, non-corrosive, smooth material that will not deteriorate when wet. Stainless steel or fiberglass paneling are recommended. Galvanized aluminum, Formica, and vinyl-covered fiberboard panels can also be used.

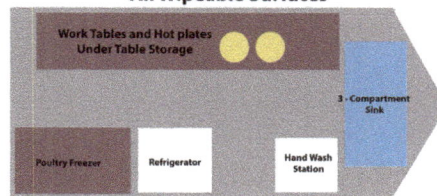

Processing Station Layout Diagrams:
Processing Kitchen Diagram

7 ft by 12 ft Trailer Vinyl lined and flooring.
All Wipeable Surfaces

Work Tables and Hot plates
Under Table Storage

Poultry Freezer

Refrigerator

Hand Wash Station

3 - Compartment Sink

Sinks should have two or three compartments, and facilities should all have access to hand washing facilities with hot and cold running water, soap, towels, and hand wash signs. Any equipment or area where food is exposed should also have shatterproof lighting overhead.

Processed poultry must be cooled to 40°F (4.4°C) within four hours (unless sold before that timeframe). Before a permit or license approval, your facility must demonstrate the availability and efficiency of your refrigeration equipment (such as refrigerators, freezers, coolers, ice chests, or insulated boxes with gel ice). A calibrated thermometer inserted into the first slaughtered bird that you can monitor will work.

Labeling requirements are based on the Federal Fair Packaging and Labeling Act of 1966. All processed foods packaged for retail sales and sold to wholesale or retail businesses must have labels. This includes processed foods sold at farmers markets, to restaurants, grocery stores, and online. Labels must properly identify the product, list the net weight, offer an ingredient statement, include food allergen labeling, contain the name and address of the packer, display a perishable shelf life pull date, and offer nutritional value.

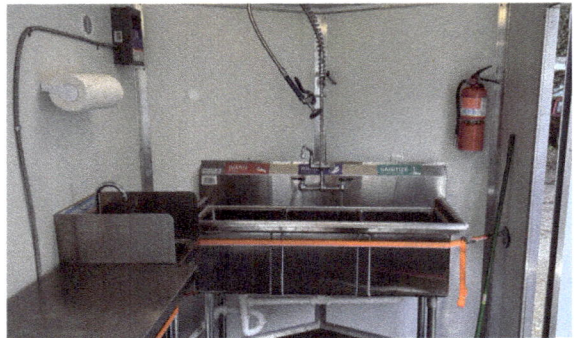

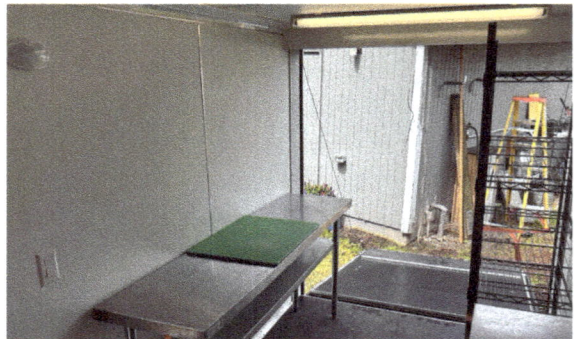

INSURANCE

Insurance coverage can be found for nearly any farm activity, but costs vary, so shop around for the insurance that best suits your needs. For small farming operations, you may be able to add coverage to your homeowner's policy. If you do direct marketing off-farm, talk with a qualified insurance agent and discuss precisely what your policy will and will not cover. Just because it says "farm" insurance, it doesn't necessarily mean that it covers barns, equipment, animals, or other farm assets. Be sure to research the details of the policy. Factor in any additional vehicle insurance needs, look into liability insurance to protect you from claims of damage or injury, and you may want to look for insurance that covers crop or revenue loss.

- Commercial general liability (CGL) insurance covers lawsuits and losses from bodily injury or property damage. This type of insurance serves the needs of many small and direct marketing farms, and coverage typically ranges from $2-5 million

- Farm insurance covers barns, rental housing, equipment, animals, and other farm assets.

- Workers' compensation insurance is required for employees or interns.

- Environmental pollution insurance covers the costs related to the clean-up of manure or pesticide contamination.

- Vendor insurance can cover farmers markets or trade show liability. Most farmers markets will require vendors to carry this type of insurance, so farms should confirm that their coverage includes business activities at markets.

- Umbrella liability coverage provides extra insurance protection over and above your existing policies, and usually carries a higher deductible.

- Poultry farm insurance protects your poultry growing facilities, laying houses, egg gathering facilities, service buildings, equipment breakage, and against production losses due to weather, pests, and other causes of loss.

There are insurance programs dedicated to diverse farms. Whole-farm revenue protection insurance for direct marketing farms is available through many insurance providers, and would cover all crops and livestock under a single policy. There are also provisions for beginning farmers and ranchers who may not yet have five years of documented revenue.

USDA DISASTER ASSISTANCE PROGRAMS

There are several federal disaster assistance programs available to farmers. These cover natural disasters such as fires or storms, which may affect livestock among other things. Some programs are oriented toward beginning farmers. The Noninsured Crop Disaster Assistance Program (NAP) is a baseline catastrophic coverage program for producers that aren't eligible for other crop insurance.

BUSINESS LICENSE REQUIREMENTS

The requirements for a business license will differ by state, province, and country. In our own personal experience, the state of Washington requires a business license for any business that grosses more than $12,000 annually (as well as any business that needs a specialty license to operate, even if you gross less than $12,000). Our state's business license application process is also used to apply for specialty licenses, obtain a country or city business license, register a trade name, hire an employee, or change a business location.

The state also requires you to define ownership structures, including LLCs (Limited Liability Company), partnerships, or corporations. You will also need to determine if you plan on having any employees within 90 days of startup to account for workers' compensation coverage.

- Specialty licenses are added to your business license, and are legally required. Common examples of specialty licenses are eggs off the farm, taxable items, pesticides, seeds grown by other farms, or even alcohol (mercifully, this doesn't affect a few well-earned after work beers).

- A Pesticide Applicator License ensures that pesticides are used safely and legally.

- A Weights and Measures License for anyone who uses scales or other measuring devices to sell products at farmers markets, farm stands, U-pick businesses, etc.

- Food safety licenses and permits are required when handling, preparing, or processing food and edible products. For poultry, your Specialty Poultry Permit or Food Processor License will cover this.

ORGANIC CERTIFICATION

Organic certification is a voluntary type of certification that a farmer may or may not choose to undertake. The USDA-certified organic seal enables consumers to easily identify products that meet national standards, and have been independently verified by an accredited third party. It is a way to differentiate your product from others, potentially enabling you to sell for higher prices. Certification can be used for all, or part of their crops, livestock, or processed food. Although we appreciate and respect those who undertake this certification, it is not one we can speak on, as we chose not to do it due to things outside of our ability to control, like neighbors using chemicals within a certain distance of the birds, etc.

8

The Business of Selling

If you're looking for the best way to sell your chickens, there are countless books dedicated on different approaches you can take. For this book, we will primarily focus on the business details of record keeping, expenses vs. profits, and how to develop ways to sell your birds to the public. At least, we will give you a snapshot of how business is conducted in our region, the Pacific Northwest of the U.S.

First and foremost, as noted in chapter seven, if you only have a Special Poultry Permit, you can only sell directly from your farm to consumers. In Washington state, it requires a Food Processor License to sell your product from anywhere other than your farm, including places like a farmers market, to hotels, restaurants, food service institutions, grocery stores, and through wholesale food distribution. You aren't allowed to sell outside of the state, and different areas have different rules. This should give you an idea of what to expect.

So! How and where do you get started selling to the public? You are producing a product and want to let people know about your product, which means that you must become an expert in marketing, advertising, and distribution for that particular product. A marketing plan must set realistic and achievable goals, and be complementary to your skills and interests. Understanding production costs, supply (yours and the competition), local demand, customer demographics, market pricing in your area, and consumers buying habits. Designing a marketing plan is an important step; as such, there are plenty of books, websites, and other resources that can help you better than we can.

Danny's Details: When we were first starting out, we went to several farmers markets and talked to people. I love discussing our passion for raising humanely treated, pasture-raised poultry. Think of it like this: you aren't selling your birds so much as you are selling yourself. We also created channels on social media where I would show snippets of our farm life, which allowed me to advertise ourselves (made the time pass faster during chores around the farm).

With my skills from my first career in broadcasting, I decided to document our life on the farm. I filmed lots of footage, then edited it and put it up on the social media platforms, including YouTube. There are lots of people out that are fascinated by farm living, or just want to have something soothing playing in the background while they work.

I initially thought I needed to create a heavily polished product like I was used to doing in TV, but as it turns out, a lot of people really liked the raw videos and longer content that was not necessarily story-driven. Day-in-the-life- videos are really popular, and don't need much polish (although I could never quite give up editing a little). Additionally, we shot tons of photos and posted on Facebook, Instagram, and other platforms.

In addition to the social media and video content, as we were just homesteading/ learning, I wrote a blog every few weeks. I also found that you shouldn't overlook traditional forms of outreach like newsletters, flyers, business cards, and a website. Most of our marketing was word of mouth, but all of the videos I made and posted along with daily farm posts on Facebooks drove a lot of folks to our farmers market booth as well.

MARKETING IS KEY

Reaching out to your potential buyers and establishing a solid market analysis is best done long before you raise your first chicken. Know the buyers in your area and what they want. Are they looking for young, eight-week-old chickens, or do they prefer a chicken raised over three months? Do they want organically raised birds, or are they fine with any method? How large is your market? How much competition is there? Go to farmers markets and talk to other booth owners. Look for other competition at the supermarkets, and find other local farms to determine the prices they are charging for whole or processed chickens. What other types of products do they sell along with their chickens? How are their signage and branding? How is the packaging of their product? Do they have flyers and brochures for the general public? Do they tell the story of their farm and how they produce their product?

The next step is to get the word out to your community that you are in business. Establish a web presence that tells your story and offers a contact point. There are many ways to do this, from a bare-bones website to something fully developed and professionally done. Use keywords such as "organic" (if appropriately licensed), "hormone-free," "lean," "free-range," and more that tell your story and help you rank well on search engines. Use what works for you and your budget.

Word of mouth is one of the foundations of marketing your birds. That includes both the product you're selling and the great service you provided. Word of mouth can start through friends, family, a church or another type of community group, draws and contests, and other similar methods. It's important to maintain a customer database detailing preferences, amounts, and contact information. Check in with your customers periodically (not exhaustively) and determine how best to serve them individually. Use surveys and suggestion boxes so you can listen to their feedback. Are there any products they want that you don't currently produce? If a customer stops buying from you, find out why.

Develop a professional and simple order form (both paper and electronic) for you and your customers. Ensure that the form displays your business name and contact information, along with the date and time the form is filled out. Provide space for the customer's name, address, phone number, and email. Ensure your form indicates how the customer will receive their order (pick up or delivery). Clearly show how the customer is paying for their product, including any deposit, taxes, and totals. Also, include the total number of pounds and number of chickens purchased. For a paper copy, provide copies for yourself and your customer. For an electronic form, keep a record in your database and email a copy to your customer.

With a website you can establish what you have for sale, how your product is produced, and take orders (including pre-sales, which is very important). You can update prices when products are available, announce upcoming products, and receive necessary public feedback about your product.

Social media sites can also help promote your efforts. Keep your smartphone or tablet handy, take pictures and videos of your work, your products, and daily life on the farm. Post those often. Re-share interesting articles, videos, and information to keep ongoing contact with your public. The use of online sales sites such as Craigslist is also a possibility.

Flyers and business cards can be a good way to get the word out about your birds. Not everyone will be interested, but it will be a physical thing you can hand someone who has questions, but may not have the time to sit down and talk. A well-made flyer can also offer good information on your farm, family, and ways you raise your chickens that might appeal to your market.

As mentioned above, pre-sales programs can take some of the risks off your shoulders by providing you with some money (deposit) upfront, and guaranteeing the delivery of a bird. Advertise your product availability, terms, and expected delivery times to your community and take orders.

A Consumer Supported Agriculture (CSA) program is another avenue to approach your market. This is a production and marketing model where a consumer buys shares of your expected harvest in advance, paid in either a lump sum or installments. It's somewhat like a pre-sales program on steroids as it allows your consumers to get a product delivered on a set schedule, and allows you to have some level of commitment and resources without the guesswork of a purely open market.

Producer Cooperatives, also known as a Marketing Cooperative, allow members who produce a similar product to sell it cooperatively. Co-ops market make up 30% of the total agriculture products in the U.S.

PICK-UP OR DELIVERY

As noted, there are different options for how you get your birds to your customer depending on your license. Suffice it to say, the more options you have, the more ways you can sell. Remember, though, that too many options can pull you in too many directions. Find the balance point that works for you.

Whether you're selling fresh or frozen, having your customers pick up your chickens directly from your farm is a convenient way for you to handle your sales in a controlled manner. You can either schedule dates and times to have the customer come to you, or set aside a single day for everyone to get their orders. Individual dates can be great for customers, but can force you to alter your entire process, including adding more butchering times per month. All at once can simplify that, but requires an entire day or two focused on butchering, packing, and storing for sale. These options become more complicated depending on whether you can only sell fresh chickens within 48 hours, or freeze them.

Many urban consumers are unfamiliar with farming and farm life, so you'll want to ensure that their visit to your farm is a positive experience. Make sure your roads are drivable and easily navigated. You might want to consider putting up signs on strategic corners and streets that point to your farm (with permission from land owners as needed). Have someone available in a roadside booth, or keep an eye out for visitors. You don't want your customers arriving at your farm, but then not be able to find you. Be clear on what form of payment you accept before they come to your farm. If you have the capabilities, I urge you to invest in a card scanner for debit and credit cards. Barring that, ensure your customers know what you will accept as payment well before they come to your farm.

If you have a delivery service available, minimize delivery times and determine how you want to stack deliveries. Keeping your chickens cool (under 40°F or 4.4°C) or frozen (under -4°F or -18°C) is important, and a small freezer or cooler with blocks of ice will help maintain the temperature (and should comply with all necessary regulations).

Ensure your packaging is clean, neat, and tidy. Presentation is important. You might consider putting frozen water bottles in with the package to keep the chicken cool, and provide several business cards with each order. Most may get thrown away, but several will also circulate among your happy customers' friends, neighbors, and family.

FARMERS MARKETS

Here's the big one: farmers markets. Farmers markets are an important component of selling your chickens to your local customers, getting the word out on your product, and reaching people that are already looking for what you are selling. These are great places to build real relationships with current and future customers. You are "their" farmer, and they will want to invest in you as a seller and person. It also allows you to set your prices mostly independent from store-bought, and capitalize on the fact that your chickens are not raised in a tiny cage.

Farmers markets are also a great place to learn the business in a low-risk, easy-to-enter environment. You will need the proper license, but once you have that, you will get to know your fellow farmers, and try different options until you find your niche with varieties, sizes, colors, prices, seasons, promotions, and more. It is also a great place to do something you love and bring healthy food directly to your community.

Most farmers markets, at least in Washington state, are non-profit organizations run by cities or, in a few cases, by private businesses. They are a great way to cut out the middleman, and allow small farms and producers to sell directly to shoppers without much overhead. Many are part of a larger strategy by local governments. Some are designed to cater to tourists and bring in businesses from nearby communities, others are part of downtown revitalization projects. They can also be run to help bolster entertainment venues, and other events.

Farmers markets are typically run by a board of directors, which you can become a part of if you want to help organize and run things. It's a valuable way for you to become involved in your community, but it's a choice that may or may not be right for you.

Be aware that most farmers markets are not a "show up and set up" format. You will need to reserve your space, get approval of your proposed sales and booth makeup, understand and adhere to all market rules, respect the established culture, pay relevant fees, and provide quality products that will represent both you and the market in the best light. Most of these guidelines should be readily available on the farmers market's official website, or barring that, you can always reach out to the market manager for all necessary steps. Some markets require all vendors to become a member of their organization, a process that needs to be renewed annually.

Booth fees can be a flat rate, a percentage of sales, or a combination of the two. Each method for paying fees can have benefits or detriments depending on your situation. The information on these fees and how they are charged should be available through the market's website or manager. Most vendor fees are collected at the end of each market day.

Be aware that not all farmers markets will be for you. Research them, and if possible, attend them before you commit. There may already be a glut of sellers that are similar to you, or you may corner the market. This can have a dramatic effect on how you set your prices, and the profits that you make. Ask the other vendors questions to see what they are and are not happy with.

MARKET BOOTHS

Undoubtedly, you will want to set up a booth that differentiates you from your competitors, but some items are a "must-have" for any successful booth. The following is a list of items that should be useful for your booth, and for any customers that stop by.

- A tent or canopy is necessary to keep the sun and rain off of you. Many quick set-up types are available, from 10x10 and larger. Be aware of the booth size requirements for any market you are attending.

- Tables or other display structures are an obvious necessity. This may be tricky for selling chickens, as you must keep your product in coolers throughout the day. Having other products for sale is generally good, so your tables do not look "empty." We got creative with this. One season it was so hot that I made a bunch of fake ice cubes out of foam, and took high-quality photos of different meat cuts in the fake ice. When it was a little cooler, we laid the meat in shallow tubs full of ice for display, rotated often, and sold what was in the ice first. GET CREATIVE!

- Banners, signage, and displays for your business and product are vital. Double check the market rules regarding what can and can't be displayed.

- Tent weights are necessary, and they must always be attached to tents or canopies. Failing to follow these guidelines can lead to serious safety and risk management issues. In Washington, a common requirement is to have at least 24 lbs anchoring each leg, and a market umbrella must have at least 50 lbs. You can use a canopy or umbrella, and it needs to be secured to grass or unpacked soil by either steel auger anchors or spiral tent stakes of at least 0.5-inch thickness, and 12-15-inch in length. Non-spiraled straight tent stakes are not acceptable, and a good idea is to make sure canopy weights are not a danger. You want your customer looking at your wares, not at their feet.

- Certified scales are a strange case. A third-party scale certifier must show up to certify where the scale is used, even though it is mobile. Rules in the industry can sometimes conflict with others, and you have to be mentally aware and prepare to be flexible when you encounter them.

- A cash box, card reader, market day sales report forms/pads, and receipt pads are all a must. You should also have a market currency "cheat sheet" with you.

- Bags for customers are important, and make sure you know your state's rules regarding paper or plastic.

- Garbage cans are helpful, and in some cases are required by the market.

- Booth amenities will save your sellers a lot of discomfort. Standing mats, rain protection, heaters for winter (with permit), and chairs are simple comforts that go a long way

- Booth supply kits should include duct tape, markers, pens, index cards, post-its, and tools.

- Business cards should be easy to access and hand out, along with, brochures, promotional material, and info about your products and business.

Think about how you can create added-value situations. One year, we teamed up with other booths to sell herbs and rubs that went well with chicken. Another time, we made soup packs that included back meat from the carcasses, along with garden herbs, onions, and tomatillos. Try out different strategies. You will be surprised how well they can work.

You'll also want to understand the rules surrounding market currency, which involves farmers markets that use special tokens, or some other form of "paid system" system in lieu of direct cash. This can be used for customers that use the Women, Infants, and Children (WIC) program, Senior Farmers Market Nutrition Program, Supplemental Nutrition Assistance Program (SNAP) market match, and market bucks or other promotional systems created by a farmers market.

RECORD KEEPING, COSTS, AND PROFITS

Profits! This is potentially one of the main reasons you are planning to raise chickens and sell them to your neighbors. It is not the only reason, as we've discussed in previous chapters, but it is a vital one if you are pursuing this as anything more than a hobby.

Before you start raising meat chickens, it's important to create a detailed plan. Estimate the number of chickens you intend to raise, the duration of the project, and the facilities required. This plan will serve as your roadmap for budgeting and tracking costs. You isolate these costs and differentiate them from the rest of the costs associated with running your farm, but it will most likely be worked into your overall budget. We won't go into the overall budgets of farms, which can vary wildly. Instead, we will focus on the costs related to raising meat chickens.

You will want to categorize your budget into different costs and categories. Startup and one-time costs should be included, but marked as non-repeating or semi-repeating. That might include items like feeders, coops, fencing, housing, and farm improvements you incorporated specifically to help establish the ability to raise your chickens.

The next category will be repeating costs that will be an ongoing part of your business. Feed costs are a major expense and subject to price changes as the birds reach different growth stages. Maintaining housing and infrastructure will impact the budget, including coops, fencing, and other items for the chickens' comfort and safety. Healthcare costs will include vaccinations, medications, supplements, and other veterinary care. Labor might include hiring help or other labor costs, and rental equipment costs can also be a substantial. The final subject would include miscellaneous expenses that may arise while raising your chickens.

Expense tracking is all about keeping meticulous records of all expenses incurred. There are many ways to do this, but I recommend you use a spreadsheet, accounting software, or even specialized farming apps to record each expense, including the date, item, quantity, unit, cost, and total cost. Sub-section these monthly and yearly so that you can keep track over time.

Revenue and income are the next step. Research and estimate the market price of chickens in your area, along with the USDA market reports for cost of birds across the country and regions. Again, keep track of this over the month and year to build your historical trends. I like keeping expenses and income on the same spreadsheet for ease of record keeping.

Regularly analyze your expenses and revenue to calculate your profits. Subtract your total expenses from your total revenue to determine your net profit. This will give you insights into the financial health of your operation.

Use the data you gather over time to make informed decisions about your poultry farming operation. This can include expanding your operation, adjusting feed types, optimizing housing, and more. Remember, accurate record-keeping and proactive financial management is essential for the success of your meat-chicken farming venture. Regularly reviewing your financial performance will help you make informed decisions, and ensure the sustainability of your operation.

9

The Road Ahead

So, we have come to the end, and I've given you all the information you need to succeed. At least, I hope I have. I want to see you succeed, because the more you succeed, the more your friends, family, and community will benefit. That is a good thing. A rising tide lifts all boats. Here are some tips and tricks I used, which I hope you find useful.

We have a family farm. All the labor we use is from family members; our kids, parents, friends, and neighbors all help out. This is a great way to enrich your personal life and the community. Remember that if you want help, you must also give help in the spirit of "barn-raising," an old farming tradition where a community would get together to build or restore someone's barn. Reach out to your neighbors and see if you can help them, and vice versa.

Don't carry long-term debt if you can avoid it. Don't get me wrong, there are circumstances where taking long-term debt can be good, but go into that with your eyes wide open. Long-term debt can pose significant burdens for interest over time. The higher the interest rate, the more it will eat into your profits and diminish the viability of your farming business. Carrying long-term debt also poses risks due to the uncertainty of agriculture. Your interest rates won't change without your action, but the volatile nature of the market can lead to sudden up or downturns regardless of what you do. Long-term debt will also result in reduced flexibility, not to mention the stress it adds to your life. Keep a close eye on your expenses, especially in the beginning when costs and profits are historically skewed.

When possible, buy old equipment and learn to fix it yourself rather than buying new. This requires a robust knowledge of maintenance and DIY techniques, but some personal education can be a viable alternative to spending money. In many ways, this can also apply to using existing buildings and infrastructure. When possible, try to construct things to be portable or easily disassembled. Try to find things that will serve multiple roles. A great re-use example is using watering troughs that no longer hold water as upcycled chick brooders.

Work to increase the health of your land. Find synergistic ways to use grazing fields and educate yourself on the best ways to do this. Be mindful of proper pasture rotation, because you can either do it well, or you can do it in a very unhealthy way that will cost you money and time. Work hard to raise the health of your land, and it will take care of you.

Ultimately, you want to run a profitable farm enterprise that provides healthy food for your family and community. This takes hard work, but it also requires an understanding of your abilities. Can you make daily efforts to provide your chickens with a healthy life? Have you considered that you will need to end the lives of your chickens respectfully? Think through these considerations before you begin, because you don't want this to come up as an issue when you are five weeks into your chicks' life.

THE (CHICKEN) TALE OF A TYPICAL SUMMER

So, you've made it to the end of our book, and you may be are wondering how the heck did two people handle 1,000 birds in three months. I'd like to layout the year for you so you have a good vision of how it might look:

January and February

If you are living solely off the farm, this is the time that you prep and plan out your year. Carry out basic farm maintenance; oil your boots, fix your tractors, read, fix old shelters, build new chicken tractors, map your fields, plantings, rotations, and take care of other necessities that need to be done.

Handle your financial responsibilities and make sure you have the money you need to finance your coming year. Do your taxes early and collect receipts. Your first paycheck will come in late June, so make sure your finances are sound. Farms have a lot of tax credits that can be taken advantage of; I was always behind the ball on this since there's always something other than taxes to do.

Coordinate your upcoming seasonal logistics before your first bird lands in the brooder. Reach out to hatcheries and feed stores, order your birds, and make sure you're able to pay for it all. Organize the selling and collection of pre-season punch cards or other types of pre-order packages. Decide what kind of combination seasonal specials that will coordinate well with your chicken sales. Come up with different value adds; we sold packages with recipes, herbs, and vegetables all together as a meal pack. Use this time to produce any additional items you will want to sell at your booth, such as fiber art or other crafts.

Build good habits from the beginning. Anything you can do in the cold months will help you immensely as you move into spring and summer. What you don't want to happen is for all those non-urgent important tasks to become urgent in the middle of the summer when you barely have enough time to breathe!

March

This is the last month before your birds start to arrive. If you haven't done it already, you need to prep and test your brooders, heat lamps, and any other equipment you will be using to raise your birds. Create a calendar with your detailed plans for the entire process, from brooders through finishing tractors. We used an online calendar program that allowed us to access it no matter where we were. June will come around fast, and you will be thankful you have a plan for cleaning, sanitizing, re-bedding, and receiving your birds. You'll also want to know what to do when it's time to move your chicks to their next stage, handle tractor rotation, slaughtering, butchering, and when it's time to sell. This is exhausting work, and when we get tired, we can miss things. A well-realized schedule will help save your sanity.

This is also the time when you need to find a dry space to start to prep for the farmers market and your booth setup. This can be an involved process, and you will want to schedule your inspector, making sure you have jumped through all of the city, county, and health authority hoops so you don't get burned by missing a detail that is required to sell your products. Checklists are important!

April

Your chicks will start to arrive this month, and continue through August/September. Ours arrived the last week in April, which meant our first butchering was the last week in June. Early to mid-month is a continuation of March to some degree, but the ground should be warming up, so finalize anything you put off until it got warmer.

May

Around this time, your birds will go from brooders and grow-out pens, and enter their pasture tractors. Your second batch of birds will arrive two weeks after the first, which should trigger the move of the first batch into the grow-out pens. By the fourth week of May, your third group should reach the brooders, the second moving to the grow-out pens, and the third will take their first forays into the pasture. Keep your grow-out pens close to the barn for easy access to heater lamps and plates, since May will still be too cold for them. As the season progresses, cold drafts will diminish, but for now, they are still a real danger. Depending on the temperature, you may even need to run a heat lamp in the tractor pasture in case your first batch isn't fully feathered yet.

Check on chicks and pullets often as you move them into their new spaces. You really want your first batches to be watched closely, as these guys will kick off your season, and in my experience spring chickens tend to be fatter if you keep them healthy until late June. (I think their heavier weights are due to the really great grass growth in the early spring.) This is your first shot at impressing your future clients, so make it count. By the end of May, you should have 300 birds on the ground.

June

A week or two into June, your fourth batch will arrive (two weeks after your third). All of your birds will move up to the next level again. The first round will go into the finishing tractor, the second will go to the pasture, the third will go to the grow-out pens, and the fourth will be in the brooder. You are now at capacity with the Misty Frog Acres system, aka raising 1,000 chickens for a farmers market (with just two people). There should be 400 birds on the ground, and you are about two weeks away from your first processing date. Your daily chores involve changing bedding, moving tractors, and removing or culling any sick birds or birds that are failing to thrive. Does it seem like a lot? Just wait until the end of the month!

You now have two weeks for final preparations in your processing and butchering spaces, and also preparation for the market. By the first week in July, your fifth round will arrive, and your first round will be ready to process and butcher (at eight weeks old). I highly suggest you stagger the arrival dates by a day or two or add a week in the middle for a slight break between birds coming in. It gives you a breathing week to get your head wrapped around the processing and butchering flow along with the raising and moving flow.

July

Full blast! The first week in July is your first butcher and market week. You will still have all your normal chores of raising the birds, but now add processing and butchering to your mid-week schedule, and market days on the weekend. This will mean long days, so prepare yourself accordingly.

If you are shooting for fresh chicken delivery at the market (i.e., never frozen), this timing is critical. The processing week should look like the following: Slaughter 100 birds on Thursday, then butcher on Friday. That involves parting out and packaging 50-75 birds, and leaving 25-50 whole. Prep and set your market booth on Saturday, and finally, on Sunday sell and tear down the market booth. We would sell fresh on Saturday and frozen on Sunday (make sure to confirm what your local health department will allow when doing this planning).

The week after that is a non-butcher week (which is really nice after an 80-100 hour butchering week). This is the time to clean and prep the processing and butchering kitchens for the next processing day. Clean and sanitize everything. You will probably still have some poultry to sell, so we turned these into our frozen-only market weekends, and depending on sales, we would just be there one day. You should be at the farmers market at least three weekends a month, but as you get more popular, you may need to reduce days at market as demand will go up and supply will stay the same.

August and September

The late summer and early fall months will look very similar to July. You will be raising birds, moving them daily, going to get feed and bedding, processing, butchering, selling, and cleaning kitchens on a two-week cycle. If you want to give yourselves an extra day or even a week, delaying the delivery of a batch of chicks can happen. Just be aware that the more down days that you put in, the longer your season becomes. That puts you at risk of going into the cold months for butchering, which is terrible.

October and November

The season will wind down and you should finish up with your last batch of birds, but it will be very busy and exhausting. One year, we decided to do fresh turkey, ready 48 hours before Thanksgiving. It was extremely popular, but we never did it again. Processing heavy birds in really cold weather after an incredibly long ranching season was just too much. This is also the time when you have the chance to sell off any remaining soup bones, feet, or other less popular cuts and clear your freezers. Typically, our last market day is generally around the end of October.

After the last market day, I strongly suggest walking out to your farm and observing the quiet chaos. Note the repairs needed, the fields covered in manure, and the hundreds of other things you worked through for the past few months. Now is your time to get everything put away and prepped for the winter. I would bring in all processing equipment to be cleaned and winterized as thoroughly as possible.

You Made It!

You have now survived a 1,000 birds-to-market summer, made new friends, made happy customers, and rescued chickens from all sorts of dangers. You have thanked 1,000 birds for their gifts, dug gut holes, laughed, cried, and are ready for a well-deserved break. You made it!

Working on a farm, even just part-time can be a daunting prospect. It isn't for everyone, and there's a steep learning curve for people that may have spent their lives in a city. But for those willing to dedicate themselves to it, those that are willing to accept that there will be days that challenge your resolve, the rewards can be incredible.

Even when I look back at all the problems we've had – the days when we didn't know if we would make it, the times when I wondered if I would physically be able to continue – I just take a minute to look at what we've accomplished and it makes it worth it. Afterall, I GET to be a farmer.

Being a small, independent farmer means you are part of a community, and possibly part of an ideal. We, and others like us, are trying to do our part to encourage sustainability and ethical practices. That's why we developed the Misty Frog Acres method for raising 1,000 chickens, and why we wrote this book. Maybe it can help you. And who knows? Maybe one day I'll be reading your book.

A FARM WEEK IN JULY

This schedule shows what your weeks will look like from the peak of your 1,000 chicken summer, all the way through late October. Good Luck!

Monday Rest Day

8:00 AM – 11:00 PM	Pasture Tractors - Move, feed and water birds. Farm Chores - Collect eggs, feed and water layers, ducks, sheep, and goats. General animal husbandry/vet as needed.
1:00 PM – 2:00 PM	Market Garden - Maintenance, weed, harvest, replant as needed.
2:00 PM – 3:00 PM	ShopTime - Tractor, vehicle, equipment maintenance, and cleaning as needed.
3:00 PM – 4:00 PM	Administrative - Marketing, bookkeeping, social media, computer work as needed.
4:00 PM – 5:00 PM	Pasture Tractors - PM Move, health check, fill feed and water.

Tuesday Farm Day

6:00 AM – 9:00 AM	Pasture Tractors - Move, feed and water birds. Farm Chores - Collect eggs, feed and water layers, ducks, sheep, and goats. General animal husbandry/vet as needed.
9:00 AM - 11:00 AM	Prep, wash and sanitize brooders for the next round of chicks, let dry in the Sun.
11:00 AM – 12:00 PM	Market Garden - Maintenance, weed, harvest, replant as needed.
1:00 PM – 2:00 PM	Administrative - Marketing, bookkeeping, social media, computer work as needed.
2:00 PM – 3:00 PM	Shop Time - Tractor, vehicle, processing equipment maintenance and cleaning as needed.
3:00 PM – 4:00 PM	Pasture Tractors - PM Move, health check, fill feed and water.
4:00 PM – 5:00 PM	Brooders - Add new bedding and turn on heaters for the next round.

Wednesday Deliveries/Pickups/Receiving

5:00 AM – 6:00 AM	Brooders - Pick up chicks at the Post Office.
6:00 AM – 7:00 AM	Brooders - Receive and monitor chicks.
7:00 AM – 9:00 AM	Pasture Tractors - Move, feed and water birds. Farm Chores - Collect eggs, feed and water layers, ducks, sheep, and goats.
9:00 AM – 11:00 AM	Pastures/Barn - moving sheep, checking fences, pasture and barn maintenance.
11:00 AM – 12:00 PM	Market Garden - Maintenance, weed, harvest, replant as needed.
12:00 PM – 4:00 PM	Logistics - Feed pickup, deliveries, all away from farm work
4:00 PM – 5:00 PM	Finish Tractors - Move to Processing vicinity
In the Evening	Pasture Tractors - PM Move, health check, fill feed and water. Kitchens - Final Clean, Sanitize slaughter and butchering kitchens.

Thursday Slaughter Day

5:00 AM – 6:00 AM	Pasture Tractors - Move, feed and water birds. Farm Chores - Collect eggs, feed and water layers, ducks, sheep, and goats.
6:00 AM – 7:00 AM	Slaughter Kitchen - Fire up scalder, charge all water lines, check for leaks, fill chill tank, check power and extension cords for power and safety.
7:00 AM – 7:00 PM	**Slaughter 100 chickens**
In the Evening	Pasture Tractors - PM Move, health check, fill feed and water.

Friday Butcher and Package Day

5:00 AM – 6:00 AM	Pasture Tractors - Move, feed and water birds. Farm Chores - Collect eggs, feed and water layers, ducks, sheep, and goats.
6:00 AM – 7:00 AM	Butchering Kitchen - Fire up heat shrink water, prep packaging materials, vacuum sealer, and bags
7:00 AM – 4:00 PM	**Butcher and Package 100 Chickens**
4:00 PM – 5:00 PM	Logistics - Go to Bank and get change for market
6:00 PM – 7:00 PM	Butchering Kitchen - Clean and sanitize butchering kitchen.
In the Evening:	Pasture Tractors - PM Move, health check, fill feed and water. Market - Prepare for AM load in and restock market items as needed.

Saturday Market Day

5:00 AM – 6:00 AM	Farm Chores - Collect eggs, feed and water layers, ducks, sheep, and goats.
6:00 AM – 7:00 AM	Market Garden - Harvest fresh veggies for the market booth
7:00 AM – 8:00 AM	Market - Load coolers for day two Market. Head to market and open booth.
8:00 AM – 9:00 AM	Market - Booth set up on location: displays, coolers, food safety requirements, etc.
9:00 AM – 3:00 PM	**Farmer's Market All Day**
3:00 PM – 5:00 PM	Market closes, leave the tent, remove valuables and take remaining goods home, button down for overnight security. Put remaining chicken in freezers.
In the Evening	Pasture Tractors - PM Move, health check, fill feed and water. Market - Prepare for AM, load in and restock market items as needed.

Sunday Market Day

6:00 AM – 7:00 AM	Farm Chores - Collect eggs, feed and water layers, ducks, sheep, and goats.
7:00 AM – 9:00 AM	Market Garden - Harvest fresh veggies for market
8:00 AM – 9:00 AM	Market - Load the coolers for day two. Head to the Market and open the booth.
9:00 AM – 3:00 PM	**Farmer's Market All Day**
10:00 AM – 5:00 PM	Market weekend ends - Breakdown booth, take remaining coolers home, unpack, initial sanitization of all coolers and scale. Put remaining chicken in freezers. Unload all market and booth items.
5:00 PM – 6:00 PM	Pasture Tractors - PM Move, health check, fill feed and water.
6:00 PM – 8:00 PM	Market Garden - Maintenance, weed, harvest, replant as needed.

10

Resources

Here are some resources for your use as needed.

CHECKLISTS

Chicken slaughter and prep checklist and bird chilling log.

√	CHICKEN SLAUGHTER PREP (It's the things we forget that cost money)	Time	Temp
	MISCELLANEOUS		
	Garbage bags		
	Ice		
	Disposable gloves		
	Bleach wipes		
	Soap/hand soap		
	Propane		
	Rain gear		
	Offal bowl and bags		
	Music		
	TOOLS		
	Knives/sharpening steel		
	Non-stick cooking spray		
	Butchers coat		
	Food safe hose		
	Extension cords		
	Bleach disinfectent spray for sprayer		
	Handwash station		
	Thermometers		
	Timer		
	PRE PROCESSING CHECKLIST		
	Hose and bleach prep areas (tables, surfaces, walls)		
	Position hoses and power cords		
	Clean scalder		
	Fill scalder		
	Dig gut hole		
	Clean plucker		
	Test plucker		
	Clean chill tanks		
	Fill chill tank with ice		
	Setup gut buckets		
	Setup hand wash		
	Clean kill cones		

Packaging checklist

	Tools
	Propane
	Sharpies
	Plastic shrink bags
	Zipties
	Small resealable bags for offal
	Vacuum seal bags
	Vacuum sealer
	Labels
	Scale
	Burners
	Large pots
	Cut gloves
	Soap
	Bleach
	Paper towels
	Part bowls

Market checklist (It's the things we forget that cost money)

√	Market Checklist (It's the things we forget that cost money)	
	Tent	
	Tent walls	
	Tent weights	
	Vinyl signage x3 (top banner, main sign, pasture poultry)	
	A-frame dry erase board	
	Tables (3-4, display dependent)	
	POS CRATE (wood display crate)	
	Jackery Battery (charged)	
	Square POS	
	Connecting cables	
	Cash box (check if the till is reset to $200)	
	Business licenses and documents	
	Markers/pens/thermometers	
	Dry Erase markers	
	Sharpies/Pens/Thermometers	
	HANDWASH CRATE (Plastic)	
	Handwash insulated cooler	
	Paper towels	
	Disposable gloves	
	Bleach wipes	
	Soap/hand soap	
	RIGGING CRATE (Plastic)	
	Wooden blocks	
	Rope, bungees, chain, s-hooks	
	Knife, pliers, tools	
	Wood Display Crate x2 (fake eggs, wire chickens, scale, table cloths, tabletop a-frames, Customer Bags	
	Multiple coolers (Chicken x2, eggs, ice)	
	Jams	
	Fiber	
	Produce	
	Merchandise	
	Produce ice packs	
	Ice	
	Broom	
	Garbage can (Garbage bags)	
	Water collection can	
	Additional display baskets	
	Local Line Orders	
	Flowers	

RECIPES

Cast Iron Roasted Chicken with Lemon Garlic Gravy (or salad dressing)

Recipes from Friends
Contributors - Joel and Kimberlee McCray

Ingredients

- 1 whole roasting chicken; giblets and neck removed
- 1 lemon, cut in half
- 1 bulb garlic, cut in half
- Salt & Pepper
- 1-2 cups Chicken Stock for gravy

1. Remove chicken from fridge, pluck any pinfeathers, and use paper towels to pat the chicken as dry as it can get, both inside and out.

2. Generously coat skin with salt and pepper. Salt and pepper the inner cavity as well, then tie up wings to the body.

3. Preheat oven to 425*.

4. When starting preheat, place dry cast iron skillet on middle rack.

5. While oven and skillet are preheating, stuff the chicken cavity, alternating ½ garlic head, ½ lemon, ½ garlic, ½ lemon.

6. Place chicken, breast side down, in the hot cast iron skillet and roast for 25-30 minutes (depending on size of chicken). It will look quite brown at this point.

7. After first roast, turn chicken over so the breast side is up. (If skin is sticking to pan, give it a few more minutes.)

8. Roast breast side up for 25-30 more minutes until fluids drain clear and/or thermometer in breast reaches 155*.

9. Remove skillet and chicken from oven. Let rest for 10 minutes before serving.

Leftover Chicken Avgolemono (Greek Lemon Chicken Soup)

Recipes from Friends
Contributors - Joel and Kimberlee McCray

Chicken Stock

- Use one chicken carcass per 3 quarts of water. You will end up with 2 quarts after boiling.
- One chicken carcass, stripped of meat and meat set aside for later
- One large white or yellow onion, cut into 8 pieces
- 2 Medium Carrots, chopped coarsely
- 4 Stalks of celery, chopped coarsely
- 1 Bay Leaf
- 4-6 Parsley Stalks (Or 1 Tsp dry parsley)
- 5 peppercorns

Combine all ingredients and simmer for about an hour. Strain all ingredients from the broth, discarding the vegetables and keeping only the liquid.

Soup Ingredients

- 1 cup orzo pasta
- 4 eggs
- 2 lemons
- Chicken (Stripped from the carcass before boiling)

Bring the chicken stock back to a simmer. Take 4 eggs and bet them well. Juice the two lemons and add the liquid to the beaten eggs. Beat vigorously until completely incorporated. Bring the stock to a roiling boil, then slowly drizzle small amounts of the egg mixture into the pot while stirring or whisking. This will create : "rags," or long tendrils of cooked eggs. Once completely added, reduce heat and add one cup of orzo to the simmering stock. Once the orzo is al dente, about 5 to 10 minutes, add the chicken bits. Adjust your seasoning with salt and pepper. (White is recommended, though black is fine)
Serve, and enjoy!

Processing Station Layout Diagrams: Chicken Slaughter Station 2021

14 foot by 7 foot covered shed roof structure

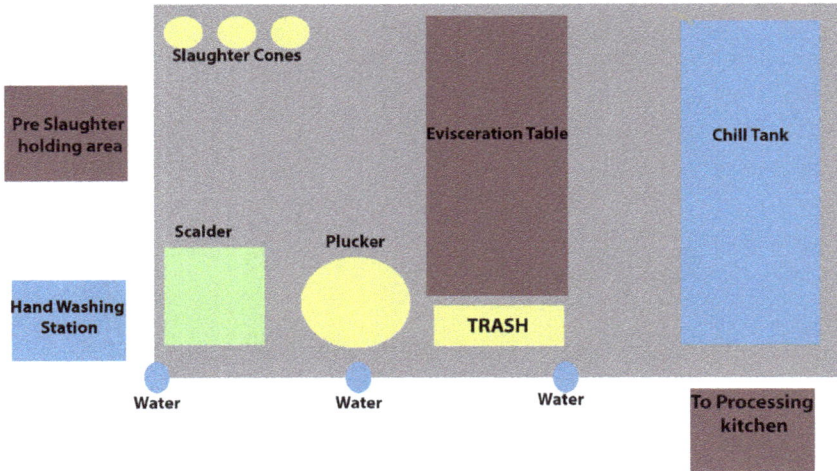

Slaughter Cones

Pre Slaughter holding area

Evisceration Table

Chill Tank

Scalder

Plucker

TRASH

Hand Washing Station

Water

Water

Water

To Processing kitchen

Processing Station Layout Diagrams: Processing Kitchen Diagram

7 ft by 12 ft Trailer Vinyl lined and flooring. All Wipeable Surfaces

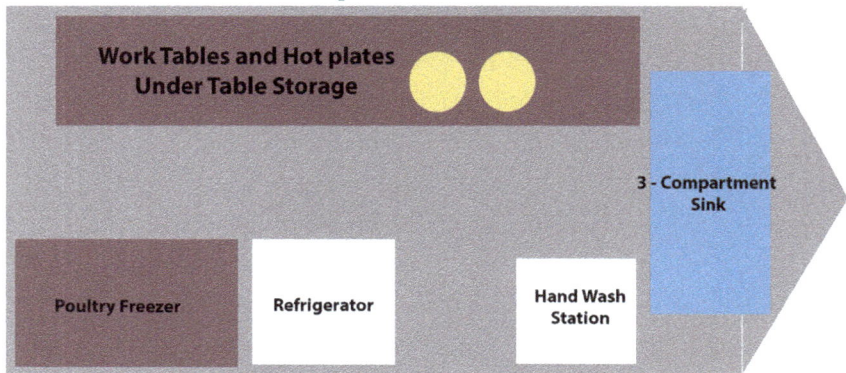

Work Tables and Hot plates Under Table Storage

3 - Compartment Sink

Poultry Freezer

Refrigerator

Hand Wash Station

Danny Rowland is an adventurer and storyteller whose curiosity for life has led him down unexpected paths. His first career as a cameraman on "Tough Guy" reality series took him to remote corners of the world, including the Arctic, where he witnessed the effects of climate change while filming polar bears in 2013. This life-changing moment spurred him to rethink his future, leading him to embrace small-scale, sustainable farming on his family's generational land.

11 years ago, Danny and his wife, Sharon, took a leap of faith and left everything behind to build Misty Frog Acres on a five-acre parcel in Battle Ground, Washington. Their dream of humane poultry farming wasn't without challenges—early on, a serious back injury nearly derailed their efforts—but their commitment never wavered. In 2021, they fully transitioned to farming full-time, providing pasture-raised, humanely treated poultry to families in the greater Portland Metro area.

A self-taught jack-of-all-trades, Danny draws on a long family tradition of inventiveness and ingenuity. From mechanical repairs and electric vehicle conversions to farm engineering, he thrives on learning new skills and pushing the limits of what's possible. As part of his commitment to sustainable farming, Danny is converting his 1975 Power King tractors into electric vehicles and installing a solar array to further reduce the farm's carbon footprint.

"How to Raise 1000 Chickens for Your Local Farmers Market" is Danny's first published work, but he's long explored creative outlets through poetry, screenwriting, and short stories. He's passionate about supporting small, eco-conscious farms that practice regenerative agriculture and foster local food networks. When he's not working on the farm, Danny enjoys rafting, relishing a good laugh by a campfire, and embracing every adventure life throws his way.